PE Mechanical Engineering
HVAC and Refrigeration Practice Exam

Second Edition

Michael R. Lindeburg, PE

PPI2PASS.COM
A **KAPLAN** COMPANY

Report Errors for This Book

PPI is grateful to every reader who notifies us of a possible error. Your feedback allows us to improve the quality and accuracy of our products. Report errata at **ppi2pass.com**.

Digital Book Notice

All digital content, regardless of delivery method, is protected by U.S. copyright laws. Access to digital content is limited to the original user/assignee and is non-transferable. PPI may, at its option, revoke access or pursue damages if a user violates copyright law or PPI's end-user license agreement.

MECHANICAL ENGINEERING HVAC AND REFRIGERATION PRACTICE EXAM
Second Edition

Current release of this edition: 2

Release History

date	edition number	revision number	update
Jun 2019	1	3	Minor cover updates.
Nov 2019	2	1	New edition.
Jun 2020	2	2	Minor corrections.

© 2020 Kaplan, Inc. All rights reserved.

All content is copyrighted by Kaplan, Inc. No part, either text or image, may be used for any purpose other than personal use. Reproduction, modification, storage in a retrieval system or retransmission, in any form or by any means, electronic, mechanical, or otherwise, for reasons other than personal use, without prior written permission from the publisher is strictly prohibited. For written permission, contact permissions@ppi2pass.com.

Printed in the United States of America.

PPI
ppi2pass.com

ISBN: 978-1-59126-659-4

Table of Contents

PREFACE AND ACKNOWLEDGMENTS v

ABOUT THE AUTHOR vii

CODES USED TO PREPARE THIS BOOK ix

INTRODUCTION xi

PRE-TEST xv

PRACTICE EXAM 1
    Practice Exam Instructions 1
    Practice Exam 3

ANSWER KEY 19

SOLUTIONS 21
    Practice Exam 21

Preface and Acknowledgments

By this time, if you are at a stage in your preparation where you are picking up *Mechanical Engineering HVAC and Refrigeration Practice Exam* and reading this preface (most people don't), then you probably have already read the prefaces in the *Mechanical Engineering Reference Manual* and *Mechanical Engineering Practice Problems*. In that case, you will have read all the clever and witty things I wrote in those prefaces, so I won't be able to reuse them here.

This book accurately reflects the NCEES Mechanical—HVAC and Refrigeration exam specifications and multiple-choice format. The format of the PE exam presents several challenges to examinees. The breadth of the subject matter covered by the exam requires that you have a firm grasp of mechanical engineering fundamentals. These rudiments are generally covered in an undergraduate curriculum. You still need to know how to select a pump, size a shaft, and humidify a room. However, exam problems enable NCEES to focus on, target, and drill down to some very specific knowledge bases. The problems within each area of emphasis require knowledge gained only through experience—a true test of your worthiness of licensure. Often, problems testing proficiency in an area can be worded as simple definition questions. For example, if you don't recognize the application of a jockey pump, you probably aren't ready to design fire protection sprinkler systems.

Needless to say, the number of problems in this practice exam is consistent with the NCEES exam, with 80 total problems. Like the exam, this book exclusively uses U.S. customary system (USCS) units. The solutions in this practice exam are also consistent in nomenclature and style with the *Mechanical Engineering Reference Manual*. Hopefully, you have already made that book part of your exam arsenal.

While *Mechanical Engineering HVAC and Refrigeration Practice Exam* reflects the NCEES exam specifications, none of the problems in this book are actual exam problems. The problems in this book have come directly from my imagination and the imagination of my colleagues, Jonathan Kweder, PE, and Bradley Heath, PE. I am grateful for their efforts.

Editing, typesetting, illustrating, and proofreading of this book continues to follow PPI's strict style guidelines for engineering publications. Managed by Grace Wong, director of editorial operations, the following top-drawer individuals have my thanks for bringing this book to life: Megan Synnestvedt, senior product manager; Meghan Finley, content specialist; Beth Christmas, project manager; Scott Rutherford, copy editor; Nikki Capra-McCaffrey, production manager; Bradley Burch, production editor; Richard Iriye, typesetter; Tom Bergstrom, production specialist; Robert Genevro and Kim Burton-Weisman, freelance proofreaders; Martin Averill, calculation checker; Sam Webster, publishing systems manager; Jeri Jump, publishing systems specialist, and Stan Info Solutions. Without their efforts, this book would read like a bunch of scribbles.

As in all of my publications, I invite your comments. If you disagree with a solution, or if you think there is a better way to do something, please let me know. You can submit errata online at the PPI website at **ppi2pass.com**.

Best wishes in your exam and subsequent career.

Michael R. Lindeburg, PE

About the Author

Michael R. Lindeburg, PE, is one of the best-known authors of engineering textbooks and references. His books and courses have influenced millions of engineers around the world. Since 1975, he has authored over 40 engineering reference and exam preparation books. He has spent thousands of hours teaching engineering to students and practicing engineers. He holds bachelor of science and master of science degrees in industrial engineering from Stanford University.

Codes Used to Prepare This Book

The documents, codes, and standards that I used to prepare this book were the most current available at the time of publication. In the absence of any other specific need, that was the best strategy.

Engineering practice is often constrained by law or contract to using codes and standards that have already been adopted or approved. However, newer codes and standards might be available. For example, the adoption of building codes by states and municipalities often lags publication of those codes by several years. By the time the 2015 codes are adopted, the 2017 codes have been released. Federal regulations are always published with future implementation dates. Contracts are signed with designs and specifications that were "best practice" at some time in the past. Nevertheless, the standards are referenced by edition, revision, or date. All of the work is governed by unambiguous standards.

All standards produced by ASME, ASHRAE, ANSI, ASTM, and similar organizations are identified by an edition, revision, or date. However, although NCEES may list "codes and standards" in its lists of Mechanical—HVAC and Refrigeration exam specifications, no editions, revisions, or dates are specified. My conclusion is that the exam is not sensitive to changes in codes, standards, regulations, or announcements in the Federal Register. This is the reason I referred to the most current documents available as I prepared this book.

Introduction

ABOUT THE PE EXAM

What Is the Format of the Exam?

The NCEES PE Mechanical exam is a computer-based test (CBT) that contains 80 multiple-choice problems given over two consecutive sessions. The problems require a variety of approaches and methodologies, and you must answer all problems in each session correctly to receive full credit. There are no optional problems.

The exam is nine hours long and includes a tutorial and an optional scheduled break. The actual time you will have to complete the exam problems is eight hours.

Exams are given year round at Pearson VUE test centers. The exam is closed book; the only reference material will be a searchable PDF copy of the *NCEES PE Mechanical Reference Handbook* provided on the computer.

What is the Typical Problem Format?

Almost all of the problems are stand-alone—that is, they are complete and independent. Problem types include traditional multiple-choice problems, as well as alternative item types (AITs). AITs include, but are not limited to

- multiple correct, which allows you to select multiple answers
- point and click, which requires you to click on a part of a graphic to answer
- drag and drop, which requires you to click on and drag items to match, sort, rank, or label
- fill in the blank, which provides a space for you to enter a response to the problem

Although AITs are a recent addition to the PE Mechanical exam and may take some getting used to, they are not inherently difficult to master. For your reference, additional AIT resources are available on the PPI Learning Hub (**ppi2pass.com**).

Traditional multiple-choice problems will have four answer options, labeled A, B, C, and D. If the four answer options are numerical, they will be displayed in increasing value. One of the answer options is correct (or "most nearly correct"). The remaining answer options will consist of three "logical distractors," the term used by NCEES to designate options that are incorrect but look plausibly correct.

WHAT SUBJECTS ARE ON THE EXAM?

The problems for each session are drawn from either the Principles or Applications knowledge areas as specified by the National Council of Examiners for Engineering and Surveying (NCEES). The exam knowledge areas are as follows.

Principles

- Basic Engineering Practice
- Thermodynamics
- Psychrometrics
- Heat Transfer
- Fluid Mechanics
- Energy/Mass Balance

Applications

- Heating/Cooling Loads
- Equipment and Components
- Systems and Components
- Supportive Knowledge

HOW TO USE THIS BOOK

Mechanical Engineering HVAC and Refrigeration Practice Exam is written for one purpose, and one purpose only: to get you ready for the NCEES PE mechanical exam. Use it along with the other PPI PE mechanical study tools to assess, review, and practice until you pass your exam.

Assess

To pinpoint the subject areas where you need more study, use the diagnostic exams on the PPI Learning Hub (**ppi2pass.com**). How you perform on these diagnostic exams will tell you which topics you need to spend more time on and which you can review more lightly.

Review

PPI offers a complete solution to help you prepare for exam day. Our mechanical engineering review courses and *Mechanical Engineering Reference Manual* offer a thorough review for the PE mechanical exam. *Mechanical Engineering Practice Problems*, *Mechanical Engineering HVAC and Refrigeration Practice Exam*, and the PPI Learning Hub quiz generator offer extensive practice in solving exam-like problems. If you don't fully understand a solution to a practice problem, you can review well-explained concepts and examples in *Mechanical Engineering Reference Manual*.

Practice

Learn to Use the NCEES PE Mechanical Reference Handbook

Download a PDF of the *NCEES PE Mechanical Reference Handbook* (*NCEES Handbook*) from the NCEES website. As you solve the problems in this book, use the *NCEES Handbook* as your reference. Although you could print out the *NCEES Handbook* and use it that way, it will be better for your preparations if you use it in PDF form on your computer. This is how you will be referring to it and searching in it during the actual exam.

A searchable electronic copy of the *NCEES Handbook* is the only reference you will be able to use during the exam, so it is critical that you get to know what it includes and how to find what you need efficiently. Even if you know how to find the equations and data you need more quickly in other references, take the time to search for them in *NCEES Handbook*. Get to know the terms and section titles used in the *NCEES Handbook* and use these as your search terms.

A step-by-step solution is provided for each problem in *Mechanical Engineering HVAC and Refrigeration Practice Exam*. In these solutions, wherever an equation, a figure, or a table value is used from the *NCEES Handbook*, the section heading from the *NCEES Handbook* is given in blue.

Getting to know the content in the *NCEES Handbook* will save you valuable time on the exam.

Using steam tables, h_1 389.0 Btu/lbm, $s_1 = 1.567$ Btu/lbm-°R, and $p_2 = 4$ psia. h_2 represents the enthalpy for a turbine that is 100% efficient. Since the turbine is isentropic, $s_1 = s_2$. Using steam tables, find the appropriate enthalpy and entropy values at state 2' where 2' = 4 psia. [**Properties of Saturated Water and Steam (Temperature) - I-P Units**]

$$h_f = 120.87 \text{ Btu/lbm}$$
$$s_f = 0.2198 \text{ Btu/lbm-°R}$$
$$h_{fg} = 1006.4 \text{ Btu/lbm}$$
$$s_{fg} = 1.6424 \text{ Btu/lbm-°R}$$

The steam quality at the turbine exhaust (state 2) for a 100% efficient turbine is found from the entropy relationship.

Properties for Two-Phase (Vapor-Liquid) Systems

$$s = s_f + x s_{fg}$$
$$x = \frac{s - s_f}{s_{fg}}$$
$$= \frac{1.567 \frac{\text{Btu}}{\text{lbm-°R}} - 0.2198 \frac{\text{Btu}}{\text{lbm-°R}}}{1.6424 \frac{\text{Btu}}{\text{lbm-°R}}}$$
$$= 0.82$$

Access the PPI Learning Hub

Although *Mechanical Engineering HVAC and Refrigeration Practice Exam*, *Mechanical Engineering Reference Manual*, and *Mechanical Engineering Practice Problems* can be used on their own, they are designed to work with the PPI Learning Hub. At the PPI Learning Hub, you can access

- a personal study plan, keyed to your exam date, to help keep you on track
- diagnostic exams to help you identify the subject areas where you are strong and where you need more review
- a quiz generator containing hundreds of additional exam-like problems that cover all knowledge areas on the PE mechanical exam
- NCEES-like, computer-based practice exams to familiarize you with the exam day experience and let you hone your time management and test-taking skills
- electronic versions of *Mechanical Engineering HVAC and Refrigeration Practice Exam*, *Mechanical Engineering Reference Manual*, and *Mechanical Engineering Practice Problems*

For more about the PPI Learning Hub, visit PPI's website at **ppi2pass.com**.

Be Thorough

Really do the work.

Time and again, customers ask us for the easiest way to pass the exam. The short answer is pass it the first time you take it. Put the time in. Take advantage of the problems provided and practice, practice, practice! Take the practice exams and time yourself so you will feel comfortable during the exam. When you are prepared you will know it. Yes, the reports in the PPI Learning Hub will agree with your conclusion but, most importantly, if you have followed the PPI study plan and done the work, it is more likely than not that you will pass the exam.

Some people think they can read a problem statement, think about it for 10 seconds, read the solution, and then say, "Yes, that's what I was thinking of, and that's what I would have done." Sadly, these people find out too late that the human brain makes many more mistakes under time pressure and that there are many ways to get messed up in solving a problem even if you understand the concepts. It may be in the use of your calculator, like using log instead of ln or forgetting to set the angle to radians instead of degrees. It may be rusty math, like forgetting exactly how to factor a polynomial. Maybe you can't find the conversion factor you need, or don't remember what joules per kilogram is in SI base units.

For real exam preparation, you'll have to spend some time with a stubby pencil. You have to make these mistakes during your exam prep so that you do not make them during the actual exam. So do the problems—all of them. Do not look at the solutions until you have sweated a little.

Take a Practice Exam

This book is a practice exam—the main issue is not how you use it, but when you use it. It is not intended to be a diagnostic tool to guide your preparation. Rather, its value is in giving you an opportunity to bring together all of your knowledge and to practice your test-taking skills. The three most important skills are (1) selection of the right subjects to study, (2) familiarity with the *NCEES Handbook*, and (3) time management. Take this practice exam within a few weeks of your actual exam. That's the only time that you will be able to focus on test-taking skills without the distraction of rusty recall.

Do not read the questions ahead of time, and do not look at the answers until you've finished. Prepare for the practice exam as you would prepare for the actual exam. Refamiliarize yourself with the *NCEES Handbook*'s layout. Check with your state's board of engineering registration for any restrictions on what materials you can bring to the exam. (The PPI website has a listing of state boards at **ppi2pass.com**.)

Read the practice exam instructions (which simulate the ones you'll receive from your exam proctor), set a timer for four hours, and answer the first 40 problems. After a one-hour break, turn to the last 40 problems in this book, set the timer, and complete the simulated afternoon session. Then, check your answers.

The problems in this book were written to emphasize the breadth of the HVAC and refrigeration mechanical engineering field. Some may seem easy and some hard. If you are unable to answer a problem, you should review that topic area in the *Mechanical Engineering Reference Manual*.

The problems are generally similar to each other in difficulty, yet a few somewhat easier problems have been included to expose you to less frequently examined topics. On the exam, only your submitted answers will be scored. No credit will be given for calculations written on scrap paper.

While some topics in this book may not appear on your exam, the concepts and problem style will be useful practice.

Once you've taken the full exam, check your answers. Evaluate your strengths and weaknesses, and select additional texts to supplement your weak areas (e.g., *Mechanical Engineering Practice Problems*). Check the PPI website for the latest in exam preparation materials at **ppi2pass.com**.

The keys to success on the exam are to know the basics and to practice solving as many problems as possible. This book will assist you with both objectives.

Pre-Test

You can use the following incomplete table to judge your preparedness. You should be able to fill in all of the missing information. (The completed table appears on the back of this page.) If you are ready for the practice exam (and, hence, for the actual exam), you will recognize them all, get most correct, and when you see the answers to the ones you missed, you'll say, "Ahh, yes." If you have to scratch your head with too many of these, then you haven't exposed yourself to enough of the subjects that are on the exam.

description	value or formula	units
acceleration of gravity, g		in/sec^2
gravitational constant, g_c	32.2	
formula for the area of a circle		ft^2
	1545	ft-lbf/lbmol-°R
	+460	°
density of water, approximate		lbm/ft^3
density of air, approximate	0.075	
specific gas constant for air	53.35	
	1.0	Btu/lbm-°R
foot-pounds per second in a horsepower		ft-lbf/hp-sec
pressure, p, in fluid with density, ρ, in lbm/ft^3, at depth, h		lbf/ft^2
cancellation and simplification of the units (A = amps; rad = radians)	A-sec^4/sec^5-rad	
specific heat of air, constant pressure		Btu/lbm-°R
common units of entropy of steam	—	
	$bh^3/12$	cm^4
molecular weight of oxygen gas		lbm/lbmol
what you add to convert $\Delta T°_F$ to $\Delta T°_R$		°
	Q/A	ft/sec
inside surface area of a hollow cylinder with length L and diameters d_i and d_o		ft^2
the value that NPSHA must be larger than		ft
	849	lbm/ft^3
what you have to multiply density in lbm/ft^3 by to get specific weight in lbf/ft^3		lbf/lbm
power dissipated by a device drawing I amps when connected to a battery of V volts		W
linear coefficient of thermal expansion for steel	6.5×10^{-6}	
formula converting degrees centigrade to degrees Celsius		
the primary SI units constituting a newton of force		N
the difference between psig and psia at sea level		psi
the volume of a mole of an ideal gas		ft^3
	1.71×10^{-9}	Btu/ft^2-hr-°R^4
universal gas constant	8314.47	
	2.31	ft/psi
shear modulus of steel		psi
conversion from rpm to rad/sec		rad-min/rev-sec
Joule's constant	778.17	

Pre-Test Answer Key

description	value or formula	units
acceleration of gravity, g	386	in/sec^2
gravitational constant, g_c	32.2	lbm-ft/lbf-sec^2
formula for the area of a circle	πr^2 or $(\pi/4)d^2$	ft^2
universal gas constant in customary U.S. units	1545	ft-lbf/lbmol-°R
what you add to $T_{°F}$ to obtain $T_{°R}$ (absolute temperature)	+460	°
density of water, approximate	62.4	lbm/ft^3
density of air, approximate	0.075	lbm/ft^3
specific gas constant for air	53.35	ft-lbf/lbm-°R
specific heat of water	1.0	Btu/lbm-°R
foot-pounds per second in a horsepower	550	ft-lbf/hp-sec
pressure, p, in fluid with density, ρ, in lbm/ft^3, at depth, h	$p = \gamma h = \rho g h / g_c$	lbf/ft^2
cancellation and simplification of the units (A = amps; rad = radians)	A-sec^4/sec^5-rad	W (watts)
specific heat of air, constant pressure	0.241	Btu/lbm-°R
common units of entropy of steam	–	Btu/lbm-°R
centroidal moment of inertia of a rectangle	$bh^3/12$	cm^4
molecular weight of oxygen gas	32	lbm/lbmol
what you add to convert $\Delta T_{°F}$ to $\Delta T_{°R}$	0	°
velocity of flow	Q/A	ft/sec
inside surface area of a hollow cylinder with length L and diameters d_i and d_o	$\pi d_i L$	ft^2
the value that NPSHA must be larger than	h_v (vapor head)	ft
density of mercury	849	lbm/ft^3
what you have to multiply density in lbm/ft^3 by to get specific weight in lbf/ft^3	g/g_c (numerically, 32.2/32.2 or 1.0)	lbf/lbm
power dissipated by a device drawing I amps when connected to a battery of V volts	IV	W
linear coefficient of thermal expansion for steel	6.5×10^{-6}	1/°F or 1/°R
formula converting degrees centigrade to degrees Celsius	°centigrade = °Celsius	The centigrade scale is obsolete.
the primary SI units constituting a newton of force	kg·m/s^2	N
the difference between psig and psia at sea level	14.7 psia (atmospheric pressure)	psi
the volume of a mole of an ideal gas	359 or 360	ft^3
Stefan-Boltzmann constant	1.71×10^{-9}	Btu/ft^2-hr-°R^4
universal gas constant	8314.47	J/kmol·K
conversion from psi to height of water	2.31	ft/psi
shear modulus of steel	11.5×10^6	psi
conversion from rpm to rad/sec	$2\pi/60$	rad-min/rev-sec
Joule's constant	778.17	ft-lbf/Btu

Practice Exam Instructions

In accordance with the rules established by your state, you may use textbooks, handbooks, bound reference materials, and any approved battery- or solar-powered, silent calculator to work this examination. However, no blank papers, writing tablets, unbound scratch paper, or loose notes are permitted. Sufficient room for scratch work is provided in the Examination Booklet. The *NCEES PE Mechanical Reference Handbook* is the only reference you are allowed to use during this exam.

You are not permitted to share or exchange materials with other examinees.

You will have eight hours for the examination: four hours to answer the first 40 questions, a one-hour lunch break, and four hours to answer the second 40 questions. Your score will be determined by the number of questions that you answer correctly. There is a total of 80 questions. All 80 questions must be worked correctly in order to receive full credit on the exam. There are no optional questions. Each question is worth 1 point. The maximum possible score for the examination is 80 points.

Partial credit is not available. No credit will be given for methodology, assumptions, or work written in your Examination Booklet.

Record all of your answers on the Answer Sheet. No credit will be given for answers marked in the Examination Booklet. Mark your answers with a no. 2 pencil. Answers marked in pen may not be graded correctly. Marks must be dark and must completely fill the bubbles. If you change an answer, be sure the old bubble is erased completely; incomplete erasures may be misinterpreted as answers.

If you finish early, check your work and make sure that you have followed all instructions. After checking your answers, you may turn in your Examination Booklet and Answer Sheet and leave the examination room. Once you leave, you will not be permitted to return to work or change your answers.

When permission has been given by your proctor, you may begin your examination.

Name: _____
      Last   First   Middle Initial

Examinee number: _____

Examination Booklet number: _____

Principles and Practice of Engineering Examination

Practice Examination

1.	A	B	C	D	28.	A	B	C	D	55.	A	B	C	D
2.	A	B	C	D	29.	A	B	C	D	56.	A	B	C	D
3.	A	B	C	D	30.	A	B	C	D	57.	A	B	C	D
4.	A	B	C	D	31.	A	B	C	D	58.	A	B	C	D
5.	A	B	C	D	32.	A	B	C	D	59.	A	B	C	D
6.	A	B	C	D	33.	A	B	C	D	60.	A	B	C	D
7.	A	B	C	D	34.	A	B	C	D	61.	A	B	C	D
8.	A	B	C	D	35.	A	B	C	D	62.	A	B	C	D
9.	A	B	C	D	36.	A	B	C	D	63.	A	B	C	D
10.	A	B	C	D	37.	A	B	C	D	64.	A	B	C	D
11.	A	B	C	D	38.	A	B	C	D	65.	A	B	C	D
12.	A	B	C	D	39.	A	B	C	D	66.	A	B	C	D
13.	A	B	C	D	40.	A	B	C	D	67.	A	B	C	D
14.	A	B	C	D	41.	A	B	C	D	68.	A	B	C	D
15.	A	B	C	D	42.	A	B	C	D	69.	A	B	C	D
16.	A	B	C	D	43.	A	B	C	D	70.	A	B	C	D
17.	A	B	C	D	44.	A	B	C	D	71.	A	B	C	D
18.	A	B	C	D	45.	A	B	C	D	72.	A B C D E F			
19.	A	B	C	D	46.	A	B	C	D	73.	A	B	C	D
20.	A	B	C	D	47.	A	B	C	D	74.	A	B	C	D
21.	A	B	C	D	48.	A	B	C	D	75.	A	B	C	D
22.	A	B	C	D	49.	A	B	C	D	76.	A	B	C	D
23.	A	B	C	D	50.	A	B	C	D	77.	A	B	C	D
24.	A	B	C	D	51.	A	B	C	D	78.	A	B	C	D
25.	A	B	C	D	52.	A	B	C	D	79.	A	B	C	D
26.	A	B	C	D	53.	A	B	C	D	80.	A	B	C	D
27.	A	B	C	D	54.	A	B	C	D					

Practice Exam

1. The water level of a 15 psig pressurized tank is 8 ft below the water level of an open tank. A pump 15 ft below the water level of the open tank delivers water through schedule-40 steel pipe to the pressurized tank. The losses due to both pipe friction and fittings are 12 ft.

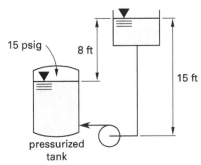

The pump must have a total dynamic head that is most nearly

(A) 19 ft water

(B) 27 ft water

(C) 39 ft water

(D) 54 ft water

2. When running at full load, a chiller has the capacity to cool 640 gpm of water from 55°F to 43°F. The rated coefficient of performance (COP) is 4.4. The heat load added by the work of the compressor that must be rejected by the cooling tower is most nearly

(A) 21 tons

(B) 73 tons

(C) 320 tons

(D) 390 tons

3. The kinematic viscosity of an SAE 10W-30 engine oil with a specific gravity of 0.88 is reported as 510 centistokes (cSt) at 37°C. The kinematic viscosity is most nearly

(A) $1.9 \times 10^{-7} \text{ m}^2/\text{sec}$

(B) $1.1 \times 10^{-7} \text{ m}^2/\text{sec}$

(C) $5.1 \times 10^{-4} \text{ m}^2/\text{sec}$

(D) $6.9 \times 10^{-3} \text{ m}^2/\text{sec}$

4. An environmental criterion for a museum requires 8000 ft³/min supply air to be delivered to a gallery space at 60°F dry-bulb temperature and 51°F dew point. If the condition of the air leaving the cooling coil is 53°F db and 52°F wb, the sensible reheat load is most nearly

(A) 22,000 Btu/hr

(B) 61,000 Btu/hr

(C) 78,000 Btu/hr

(D) 84,000 Btu/hr

5. A fluid flows through a tube at a rate of 3×10^{-5} ft³/sec. The tube is 3.4 ft long and has an internal diameter of 0.018 in. The flow is viscous with a viscosity of 26.37×10^{-6} lbf-sec/ft², incompressible, steady, and laminar. The predicted pressure drop across the length of the tube is most nearly

(A) 11,000 lbf/ft²

(B) 15,000 lbf/ft²

(C) 22,000 lbf/ft²

(D) 26,000 lbf/ft²

6. Investors are evaluating an office building that has a purchase price of $525,000. The investors expect to sell the building for $700,000 after 7 years. During the time that the building is held, the building will be depreciated using straight line depreciation, an estimated life of 15 years, and an (assumed) salvage value of zero. (The full purchase price will be depreciated.) The effective tax rate of the investors is 40%.

During ownership, the estimated expenses will be $25,000 per year, which will be offset by rental income of $45,000 for the first 3 years and $75,000 for the next 4 years.

Neglect the benefit of a favorable capital gains tax rate on the gain above $525,000. The after-tax rate of return on this investment will be most nearly

(A) 2.1%

(B) 7.0%

(C) 7.9%

(D) 12%

7. A 20 in diameter suction line carries water at 7000 gpm. The water enters a pump at 100 psi and is pumped up a 30 ft high incline through a 12 in diameter discharge line. If the pressure at the top of the incline is 500 psi and the head loss through the 30 ft section is 10 ft, the horsepower that is supplied by the pump is most nearly

(A) 140 hp

(B) 750 hp

(C) 1200 hp

(D) 1700 hp

8. Windows to be installed in a building are specified with glass having an R-value of 1.5 hr-ft^2-°F/Btu. Outdoor temperatures as low as -40°F are experienced in the winter, and the indoor space is to be heated to 70°F. Assuming indoor and outdoor window film coefficients of 1.46 Btu/hr-ft^2-°F and 6.0 Btu/hr-ft^2-°F, respectively, the maximum relative humidity that can be maintained without experiencing condensation on the inside of the glass is most nearly

(A) 31%

(B) 34%

(C) 37%

(D) 42%

9. Air with a mass of 3 lbm is expanded polytropically behind a piston in a cylinder from an initial condition of 120 psia and 100°F to a final condition of 40 psia. Using a polytropic exponent of 1.6, the work done by the closed system per unit mass is most nearly

(A) -1.7×10^4 ft-lbf/lbm

(B) -5.6×10^3 ft-lbf/lbm

(C) 5.6×10^3 ft-lbf/lbm

(D) 1.7×10^4 ft-lbf/lbm

10. A heat exchanger pump receives water at 140°F and returns it at 95°F. The water is received and returned through nominal 3 in copper tubing. The tubing has an outside diameter of 3.5 in, an inside diameter of 3.062 in, a cross-sectional area of 7.370 in^2, and an electrical resistivity of 10 Ω-cmil/ft. A circular mil (cmil) is the area of a circle with a diameter of 0.001 in. The pump is driven by a one-phase, 110 V AC, 5 hp motor that draws an average (RMS) current of 35 A. Copper tubing is placed inside PVC pipe for electrical insulation. The copper tubing carries the electrical current that will power the pump motor. Two wires (pipes) are needed to complete the circuit. The voltage drop is not to exceed 0.6 V. The maximum distance that the pump and motor can be located from a 110 V source is most nearly

(A) 870 ft

(B) 2500 ft

(C) 4900 ft

(D) 9000 ft

11. A 2 hp motor is used to stir a tank containing 60 lbm of water for 15 min. Assuming the process occurs at constant volume, the maximum possible rise in temperature is most nearly

(A) 1.4°F

(B) 21°F

(C) 58°F

(D) 130°F

12. The rectangular wall of a furnace is made from 3 in fire-clay brick ($k = 0.58$ Btu-ft/hr-ft^2-°F) surrounded by 0.25 in of steel ($k = 26$ Btu-ft/hr-ft^2-°F) on the outside. Mild steel bolts with a diameter of 0.25 in connect the steel to the brick. The furnace is surrounded by 70°F air with a convection coefficient of 1.65 Btu/hr-ft^2-°F, while the inner surface of the brick is held constant at 1000°F. Disregard conductance through the bolts. The outside surface temperature of the steel is most nearly

(A) 160°F

(B) 470°F

(C) 610°F

(D) 760°F

13. Water is used in a parallel flow tube-and-shell heat exchanger to cool a 95% ethyl alcohol solution ($c_p = 0.9$ Btu/lbm-°F). The heat exchanger is designed using 1 in outer diameter tubing. The water enters the heat exchanger at a rate of 40,000 lbm/hr at 55°F. The ethyl alcohol solution enters the heat exchanger at a rate of 45,000 lbm/hr at 160°F and is cooled to 110°F. The overall coefficient of heat transfer based on the outer tube area is 75 Btu/hr-ft²-°F. The heat transfer surface area of the heat exchanger is most nearly

(A) 520 ft²

(B) 850 ft²

(C) 920 ft²

(D) 1100 ft²

14. A steam turbine operates as a component of a Rankine cycle. Steam is supplied to the turbine at 1000 psia and 800°F. The turbine exhausts at 4 psia. The expansion is not reversible, and the exhaust is vapor at 100% quality. The thermal efficiency of the turbine is most nearly

(A) 59%

(B) 76%

(C) 85%

(D) 100%

15. The gravimetric air-to-fuel ratio of methane burned with 125% theoretical air is most nearly

(A) 13:1

(B) 17:1

(C) 22:1

(D) 35:1

16. A solid copper sphere at 852°F is dropped into a large tank of oil at 167°F. The sphere has a diameter of 7.88 inches and a thermal conductivity of 2628 Btu-in/(hr-ft²-°F). The average convective heat transfer coefficient is 155 Btu/(hr-ft²-°F) and the oil is stirred uniformly at all times. Most nearly, how long after submersion does the sphere reach a temperature of 392°F?

(A) 5.3 sec

(B) 140 sec

(C) 910 sec

(D) 4700 sec

17. A refrigeration system using refrigerant-134a (R134a) operates at 33 psia on the low-pressure side and 100 psia on the high-pressure side. The 255 ton system delivers refrigerant vapor with 20°F superheat to the compressor. If the refrigerant on the high-pressure side is cooled to saturated liquid before expansion, the needed refrigerant mass flow rate is most nearly

(A) 43,000 lbm/hr

(B) 48,000 lbm/hr

(C) 51,000 lbm/hr

(D) 55,000 lbm/hr

18. Recirculated air from a room conditioned to 75°F and 50% relative humidity is mixed with outdoor air at 90°F dry-bulb temperature and 75°F wet-bulb temperature. The air mixture then passes through a cooling coil and fan for redistribution to the room. The flow rate of the recirculated air is 7000 ft³/min, and the flow rate of the outdoor air is 2300 ft³/min. The equation for determining the total humidity ratio is

$$w_{\text{mixed}} = w_{\text{return}} + \left(\frac{Q_{\text{outside}}}{Q_{\text{return}} + Q_{\text{outside}}}\right)(w_{\text{outside}} - w_{\text{return}})$$

The total humidity ratio of air entering the coil is most nearly

(A) 0.0093 lbm moisture/lbm dry air

(B) 0.011 lbm moisture/lbm dry air

(C) 0.014 lbm moisture/lbm dry air

(D) 0.015 lbm moisture/lbm dry air

19. A centrifugal pump delivers 80°F water from a lake to a storage reservoir. The lake and the reservoir are located near sea level. The reservoir surface is 17 ft above the lake surface. The pump is located near the surface of the storage reservoir. and has a net positive suction head requirement of 12 ft water. Friction and fitting losses are 3.0 ft water. The net positive suction head available for the pump is most nearly

(A) 13 ft water

(B) 14 ft water

(C) 16 ft water

(D) 20 ft water

20. A centrifugal pump is driven at 1300 rpm by a 10 hp motor and delivers 250 gpm of 85°F water against water head of 75 ft. Pump efficiency is initially 65% and does not vary appreciably. The maximum flow rate of the pump is most nearly

(A) 280 gpm
(B) 470 gpm
(C) 520 gpm
(D) 650 gpm

21. An air conditioning system is designed for a 40,000 ft^2 casino that operates 24 hr/day. The casino is inside a larger building and has no windows or exterior walls exposed to direct sun. The lighting system requires 3.75 W/ft^2. The casino's equipment requires an additional 80 kW. Steam tables for an all-day buffet contribute an additional 50,000 Btu/hr of latent load. The maximum density for patrons is 120 people/1000 ft^2. Space conditions are maintained at 73°F and 45% relative humidity. The total space cooling load due to internal heat gain is most nearly

(A) 890,000 Btu/hr
(B) 1,900,000 Btu/hr
(C) 2,400,000 Btu/hr
(D) 3,000,000 Btu/hr

22. An office space measures 12 ft × 12 ft and has a ceiling height of 9 ft. The cooling load is 3000 Btu/hr. A single ceiling-mounted diffuser is installed at the center of the ceiling. For good occupant comfort and a maximum air distribution performance index, the throw of the diffuser should be most nearly

(A) 5.0 ft
(B) 6.0 ft
(C) 10 ft
(D) 12 ft

23. The air-conditioned interior of a car is at 70°F and 40% relative humidity when the car is parked. The car sits in a driveway overnight, causing the interior temperature to drop to 50°F, bringing the interior's relative humidity to most nearly

(A) 40%
(B) 60%
(C) 80%
(D) 100%

24. A fan supplies 4500 ft^3/min of air at 68°F through a 240 ft long rectangular duct. The duct dimensions are 18 in × 24 in. If the duct has a friction factor of 0.016 and a roughness of 0.0003 ft, the total pressure drop due to friction is most nearly

(A) 0.2 in water
(B) 0.3 in water
(C) 0.5 in water
(D) 0.6 in water

25. The water pipes in a vented crawl space are uninsulated. The space shares 320 ft^2 of exterior wall with the outdoors and 645 ft^2 of floor with a space above that is heated to 72°F. The overall heat transfer coefficients (U-factors) for the crawl space wall and floor above are

$$U_{floor} = 0.05 \text{ Btu/hr-ft}^2\text{-°F}$$
$$U_{wall} = 0.15 \text{ Btu/hr-ft}^2\text{-°F}$$

Outdoor air at 900 ft^3/hr infiltrates the crawl space through the vents. The pipes are safe from freezing down to an outdoor temperature of most nearly

(A) −10°F
(B) 2°F
(C) 10°F
(D) 30°F

26. Water enters an evaporative cooling tower as a saturated liquid at 180°F. 8% of the water evaporates as saturated vapor. After this process, the temperature of the remaining water is most nearly

(A) 81°F
(B) 86°F
(C) 94°F
(D) 100°F

27. The water surface in a well is 90 ft below a house. A storage tank with its water surface 50 ft above the house provides gravity water flow to the house. A submersible pump is capable of delivering 6 gpm from the well to the tank. The system is powered directly by a solar-electric array that has an overall efficiency of 12% and regularly intercepts solar radiation of a magnitude of at least 28 W/ft^2 on a cloudless day. The pump efficiency is 60%, and piping has been oversized to such an extent that friction and fitting losses are negligible. The minimum required area for the solar array operating on a cloudless day is most nearly

(A) 6 ft^2

(B) 20 ft^2

(C) 50 ft^2

(D) 80 ft^2

28. To reduce the load on a chiller plant, an air washer recirculates water that is at 57°F and uses evaporation to precool outdoor air at a rate of 20,000 ft^3/min. The outdoor air is introduced at 92°F dry-bulb temperature and 57°F wet-bulb temperature. The density of the air is 0.075 lbm/ft^3. The saturation efficiency of the process is 84%. The cooling requirement reduction is most nearly

(A) 45 tons

(B) 49 tons

(C) 53 tons

(D) 63 tons

29. An electric unit heater is needed to heat a room on the second floor of a three-story building. The room has no windows. The 800 ft^2 external wall is made up of 2 in of polystyrene rigid insulation sandwiched between 8 in brick having a resistance value of 0.10 (hr-ft^2-°F)/Btu-in and 5/8 in gypsum board. The density of polystyrene is 1.25 lbm/ft^3. The outdoor winter design temperature is 10°F, and the room is to be maintained at 74°F. Using a 25% safety factor, the required heating capacity for the unit heater is most nearly

(A) 1.5 kW

(B) 1.9 kW

(C) 2.4 kW

(D) 2.8 kW

30. Air at 80°F is introduced to a 3 ft × 4 ft (face-area dimensions) cooling coil at a velocity of 450 ft/min and is cooled to 56°F. The coil cooling process has a sensible heat ratio (SHR) of 0.70. The total coil capacity is most nearly

(A) 120,000 Btu/hr

(B) 140,000 Btu/hr

(C) 180,000 Btu/hr

(D) 200,000 Btu/hr

31. In a commercial air conditioning system, 2530 gpm of 83°F condenser water are supplied to an 840 ton chiller for heat rejection and are returned to the cooling tower at 92°F. The coefficient of performance of the chiller is most nearly

(A) 4

(B) 5

(C) 6

(D) 8

32. A 6 ft^3 gas bottle holds 5 lbm of compressed nitrogen gas at 100°F. Gas is released until the bottle pressure reaches 150 lbf/in^2. The mass of gas released is most nearly

(A) 0.030 lbm

(B) 0.80 lbm

(C) 1.7 lbm

(D) 4.2 lbm

33. (*Drag and drop*) Order the following list of impeller designs from highest to lowest in terms of efficiency.

impeller designs
backward curved
forward curved
backward inclined
airfoil

(A)

impeller designs ranked by efficiency from highest to lowest
backward curved
backward inclined
airfoil
forward curved

(B)

impeller designs ranked by efficiency from highest to lowest
airfoil
forward curved
backward curved
backward inclined

(C)

impeller designs ranked by efficiency from highest to lowest
backward curved
airfoil
backward inclined
forward curved

(D)

impeller designs ranked by efficiency from highest to lowest
airfoil
backward curved
backward inclined
forward curved

34. A store in a shopping mall is to be maintained at 75°F and 45% relative humidity with supply air at 55°F and 30% relative humidity. The space cooling load is 73,000 Btu/hr sensible and 26,000 Btu/hr latent at outdoor design conditions of 94°F dry-bulb temperature and 72°F wet-bulb temperature. The ventilation requirement is 850 ft³/min. The coil load due to the ventilation air is most nearly

(A) 30,000 Btu/hr
(B) 60,000 Btu/hr
(C) 80,000 Btu/hr
(D) 100,000 Btu/hr

35. The reheat coil in a variable air volume terminal box is being replaced. The maximum airflow capacity of the box is 2400 ft³/min. A minimum stop setting of 30% (of the maximum flow) has been established to maintain the required ventilation when cooling loads are at a minimum. The supply air temperature for the building system is reset with respect to the outside air temperature, according to the graph shown.

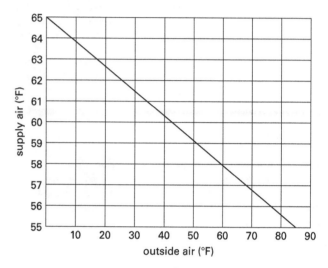

During the winter, the outdoor design temperature of 10°F and the indoor space temperature of 72°F result in a space heat loss of 45,000 Btu/hr. The minimum capacity of the reheat coil is most nearly

(A) 45,000 Btu/hr
(B) 47,000 Btu/hr
(C) 51,000 Btu/hr
(D) 66,000 Btu/hr

36. In a process, valves at each end of an 80 ft section of 2 in schedule-40 pipe are closed instantaneously, and the saturated steam (at atmospheric pressure) in the pipe begins to cool. The average heat transfer over the cooling period is 6200 Btu/hr. There is no heat loss through the valves at the ends of the pipe. The time it takes for the steam to cool to an ambient temperature of 60°F is most nearly

(A) 0.01 min
(B) 1 min
(C) 4 min
(D) 10 min

37. A simple Rankine cycle operates between superheated steam entering a turbine at 1200°F and 700 psia and entering a pump at 2 psia. The cycle's maximum possible efficiency is most nearly

(A) 27%

(B) 31%

(C) 39%

(D) 43%

38. A single-stage chiller circulates 117,000 lbm/hr of refrigerant-22 (R-22) and operates with a 90°F condensing temperature and 10°F evaporating temperature. Saturated refrigerant vapor enters the compressor with no superheat, and saturated liquid refrigerant leaves the condenser with no subcooling. Heat is rejected to condenser water that enters the condenser at 85°F and leaves at 95°F. The rated coefficient of performance is 5.5 under these conditions. The condenser water flow needed for heat rejection is most nearly

(A) 560 gpm

(B) 1600 gpm

(C) 1900 gpm

(D) 2200 gpm

39. Schedule-40 pipe is needed to transport 2200 gal/min of water through a municipal water supply system. The recommended velocity for water flowing through a municipal pipeline is 5 ft/sec. The minimum pipe diameter needed is most nearly

(A) 10 in

(B) 12 in

(C) 14 in

(D) 16 in

40. Recirculated air at 7000 ft^3/min is pulled from a room conditioned to 75°F and 50% relative humidity. The recirculated air is mixed with 2300 ft^3/min of outdoor air at 90°F dry-bulb temperature and 75°F wet-bulb temperature prior to passing through a cooling coil and a centrifugal fan for distribution back to the room. The coil has a total cooling capacity of 300,000 Btu/hr and a sensible cooling capacity of 235,000 Btu/hr. The temperature of the air leaving the coil is most nearly

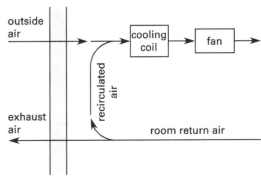

(A) 49°F

(B) 55°F

(C) 64°F

(D) 90°F

41. (*Drag and drop*) Consider the following psychrometric process diagram.

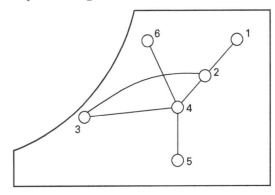

In the table shown, match the process on the left with the correct description on the right.

process	description
1-2	dehumidification
3-4	evaporative cooling
4-6	sensible and latent cooling
4-5	sensible and latent heating

(A) process 1-2: evaporative cooling, process 3-4: sensible and latent cooling, process 4-6: dehumidification, process 4-5: sensible and latent heating

(B) process 1-2: dehumidification, process 3-4: sensible and latent cooling, process 4-6: sensible and latent heating, process 4-5: evaporative cooling

(C) process 1-2: sensible and latent heating, process 3-4: sensible and latent cooling, process 4-6: evaporative cooling, process 4-5: dehumidification

(D) process 1-2: sensible and latent cooling, process 3-4: sensible and latent heating, process 4-6: evaporative cooling, process 4-5: dehumidification

42. The chiller system in a 1600 ft² (floor area) refrigeration machinery room contains 150 lbm of refrigerant. The room has 1.5 ft² of equivalent leakage area to the outside and an estimated heat gain of 15,000 Btu/hr. The design outside air temperature is 90°F. According to *Uniform Mechanical Code*, refrigeration machinery rooms must be provided with dedicated mechanical exhaust with the capacity to achieve all of the following.

I. continuous maintenance of the room at 0.05 in water negative relative to adjacent spaces, Δp, with equivalent leakage area, A_e, in ft², as calculated by

$$Q_1 = 2610 A_e \sqrt{\Delta p}$$

II. continuous airflow of 0.5 ft³/min per gross ft², A_{gf}, within the room as calculated by

$$Q_2 = 0.5 A_{gf}$$

III. temperature rise, ΔT, within the refrigeration machinery room due to heat dissipation, q, in Btu/hr to a maximum of 104°F, limited by

$$Q_3 = \frac{\sum q}{1.08 \Delta T}$$

IV. emergency purge of escaping refrigerant mass m, in lbm, as calculated by

$$Q_4 = 100 \sqrt{m}$$

The dedicated mechanical exhaust required for the room is most nearly

(A) 800 ft³/min
(B) 900 ft³/min
(C) 1000 ft³/min
(D) 1200 ft³/min

43. A water-cooled condenser heats cooling water from 65°F to 95°F. Saturated water vapor at a mass flow rate of 2200 lbm/hr enters the condenser at 4 psia and exits as saturated liquid. The mass flow rate of the exiting cooling water is most nearly

(A) 34,000 lbm/hr
(B) 66,000 lbm/hr
(C) 74,000 lbm/hr
(D) 83,000 lbm/hr

44. Underground pipes supply water at 120°F. Heat is transferred from the pipes to the surrounding soil. The flow rate of the water is controlled to maintain a constant rate of heat transfer to the soil, and the soil temperature is 55°F. During the previous summer, soil measurements showed an average moisture content of 8% and a thermal conductivity of 12.6 Btu-in/hr-ft²-°F. The following spring, soil measurements show a moisture content of 30%. The thermal conductivity of water at 55°F is 4.06 Btu-in/hr-ft²-°F. In order to maintain a soil temperature of 55°F, the

(A) flow rate of the water must be increased
(B) flow rate of the water must be decreased
(C) temperature of the water exiting the pipe must be increased
(D) temperature of the water entering the pipe must be decreased

45. A sun room has an exposed 6 in concrete floor slab measuring 20 ft × 30 ft. By nightfall, the room is heated passively by solar radiation to an average temperature of 82°F. The density and specific heat of the concrete is 140 lbm/ft³ and 0.22 Btu/lbm-°F, respectively. The room thermostat setting is kept at 60°F. Assume a floor slab convective heat transfer coefficient of 16.8 Btu/hr-ft²-°F. The average temperature change of the slab after 1 hr of cooling is most nearly

(A) 11°F
(B) 12°F
(C) 14°F
(D) 15°F

possible connected load.

To satisfy a hot-water demand of 1 hr, the recommended volume of the storage tank is most nearly

(A) 520 gal

(B) 650 gal

(C) 1700 gal

(D) 2200 gal

47. The specific heat for air at constant pressure is 0.240 Btu/lbm-°R. The specific heat for air at constant volume is 0.171 Btu/lbm-°R. The work needed to isentropically compress 10 lbm of air at atmospheric pressure and a temperature of 65°F to a pressure of 720 psia is most nearly

(A) 590 Btu

(B) 820 Btu

(C) 1600 Btu

(D) 1800 Btu

48. An east-facing vertical window at a latitude of 40 degrees north has an area of 12 ft^2. The solar heat gain coefficient for the window is 0.87. The overall heat transfer coefficient is 1.2 Btu/hr-ft^2-°F. The table shown gives the incident total irradiance for 40 degrees north.

0800	31	155	215	153	30	28	28	28	1600
0900	32	113	191	160	44	33	31	31	1500
1000	34	64	146	149	70	35	35	36	1400
1100	37	41	80	115	89	40	39	37	1300
1200	37	37	42	73	96	73	40	37	1200
half day total		252	735	1037	819	314	203	187	186	

On a day in May at solar time 0800, the indoor temperature is 75°F, and the outdoor temperature is 42°F. The total instantaneous heat gain for the window is most nearly

(A) 480 Btu/hr

(B) 1800 Btu/hr

(C) 2300 Btu/hr

(D) 2800 Btu/hr

49. A building is located in a part of the country where the annual heating season is 23 weeks. The building is occupied from 8:00 a.m. to 6:00 p.m., Monday through Friday. The building owner pays $0.30 per therm for gas heating (which includes the costs of fans, pumps, etc.), and the gas furnace efficiency is 77%. The building has the following characteristics.

internal volume	600,000 ft³
inside temperature	70°F
infiltration rate	$0.018 \frac{\text{Btu}}{°F} Q V \Delta T$
ventilation rate	
occupied	1 air change per hour
unoccupied	0.5 air change per hour
walls	
area	10,000 ft²
U-value	0.15 Btu/hr-ft²-°F
windows	
area	2500 ft²
U-value	1.10 Btu/hr-ft²-°F
roof	
area	25,000 ft²
U-value	0.06 Btu/hr-ft²-°F

In the past, the building owner kept the interior temperature fixed at 70°F. What will be the approximate annual savings if the building owner installs an automatic setback thermostat to reduce the interior temperature from 70°F to 58°F during unoccupied time?

(A) $930

(B) $1200

(C) $1400

(D) $1600

50. An existing home contains 67 incandescent light bulbs. The table shown lists the properties of these light bulbs.

number of light bulbs	wattage	lighting use factor	lighting special allowance factor
43	100	1	0.97
14	60	1	0.95
10	40	1	0.90

The homeowner replaces the incandescent light bulbs with LED light bulbs in order to save on energy costs. These light bulbs operate an average of 6 hr/day, and the cost of electricity is $0.09/kWh.

number of light bulbs	wattage	lighting use factor	lighting special allowance factor
43	14	1	0.97
14	8.5	1	0.95
10	4.5	1	0.90

During the heating season (32 weeks), the homeowner discovers his natural gas furnace is operating more often even though this heating season is similar to past heating seasons. The furnace is 80% efficient, and the cost of natural gas is $1.20 per therm. The money that will be saved by switching to LED light bulbs during the heating season is most nearly

(A) $120

(B) $260

(C) $350

(D) $410

51. A property owner has installed an air-cooled chiller on the side of his building, as shown.

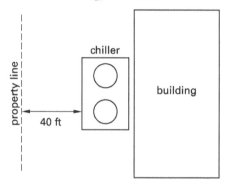

plan view

The neighboring property owner has complained that the unit is too noisy, particularly at night. The local municipal code requires that the sound pressure level be no greater than 52 dB at the property line. The municipality is sending an engineer to analyze the sound from the air-cooled chiller. The method that is most likely to be used is

(A) room noise criteria (RNC)

(B) noise criteria (NC)

(C) balanced noise criteria (NCB)

(D) A-weighted decibels (dBA)

52. An old fire sprinkler system uses 1 in steel pipe with a roughness coefficient of 100. The sprinklers are located 10 ft apart. The minimum pressure at any sprinkler is 10 psig. All sprinklers have standard ½ in orifices with orifice coefficients of 0.75. When the sprinkler system is activated, the last three sprinkler heads on a branch line are fully open simultaneously. Disregarding velocity pressure, the discharge from the second sprinkler head from the end is most nearly

(A) 11 gpm

(B) 15 gpm

(C) 19 gpm

(D) 23 gpm

thermal conductivity   29 Btu/hr-ft-°F

Saturated steam at atmospheric pressure flows through the pipe at a rate high enough to prevent substantial condensation. The average inside heat transfer coefficient is 1500 Btu/hr-ft²-°F. The average outside heat transfer coefficient of the bare pipe in still air is 2.0 Btu/hr-ft²-°F. The air in the plant is at 60°F and 14.7 psia and is normally still. The pipe temperature is too low to consider the effects of radiation. The rate of heat loss from the bare pipe is most nearly

(A) 14,000 Btu/hr
(B) 18,000 Btu/hr
(C) 19,000 Btu/hr
(D) 21,000 Btu/hr

55. A homeowner plans to have a contractor install a standard furnace with an efficiency of 78%. It is estimated that the furnace will consume 1500 therms of natural gas per year. The contractor has offered to install a high-efficiency condensing furnace with an efficiency of 92% for an additional $800. Assuming a constant cost of natural gas of $0.85 per therm, the simple payback period of this additional investment is most nearly

(A) 2 yr
(B) 4 yr
(C) 5 yr
(D) 9 yr

pressure of the water vapor in the chamber is

(A) 0.20 psia
(B) 0.55 psia
(C) 1.1 psia
(D) 2.1 psia

58. A refrigerant at –5°F passes through a thin-walled cylindrical tube with a diameter of 5/8 in. Ambient air passing over the pipe is cooled such that the heat extraction rate per foot of pipe is 25 W/ft. Under steady-state conditions, the convection heat transfer coefficient is 5 W/ft²-°F. Conductive and radiant heat transfer are negligible. Most nearly, the temperature of the ambient air is

(A) 26°F
(B) 31°F
(C) 36°F
(D) 41°F

59. Dry air flows at 30 lbm/sec into a long, insulated channel that contains a pool of water. Liquid water flowing in at point 2 is introduced as make-up water to match the evaporation rate. The dry air enters at point 1 at 80°F and atmospheric pressure, and it exits at point 3 saturated at 110°F. Most nearly, the rate of make-up water is

(A) 7.6 gpm

(B) 9.1 gpm

(C) 11 gpm

(D) 13 gpm

60. Water at 65°F flows through the pipe system shown at 48 gpm. The pipe is schedule 40 steel pipe with a nominal diameter of 0.75 in. The loss coefficients for the 90° elbows and the 45° elbows are 0.9 and 0.42, respectively.

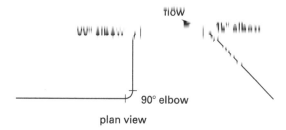

plan view

The sum of the minor losses for the system is most nearly

(A) 29 ft

(B) 33 ft

(C) 38 ft

(D) 44 ft

61. An air conditioning system takes in outdoor air at 40°F and 20% relative humidity at 6000 cfm. The air is sensibly heated to 70°F, followed by humidification with steam to a final condition of 90°F dry-bulb temperature and 71°F wet-bulb temperature. All processes are at atmospheric pressure. The needed flow rate of hot steam is most nearly

(A) 0.62 ft^3/sec

(B) 1.1 ft^3/sec

(C) 2.3 ft^3/sec

(D) 3.6 ft^3/sec

62. The pumping curve shown is for a variable speed 3450 rpm water pump. The system (or operating) curve for an application that will use this pump is superimposed on the pumping curve. At maximum design load, 600 gpm at 85 ft of head is needed.

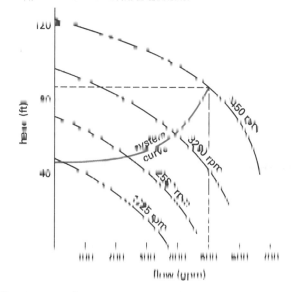

During an off-peak condition, only one-half the design flow is needed. Most nearly, the pump speed should then be reduced to

(A) 1400 rpm

(B) 1700 rpm

(C) 2600 rpm

(D) 3300 rpm

63. A scuba diver carries an air tank holding 80 ft^3 of air at 70°F and atmospheric pressure. The cylindrical air tank has an inner diameter of 7.25 in and a height of 30 in. The diver is performing moderate work to repair an underwater vessel. The tank is considered empty once it reaches 700 psig. Based on the diver's level of exertion, the number of hours of air in the tank is most nearly

(A) 10 hr

(B) 14 hr

(C) 23 hr

(D) 45 hr

(B) 5.0 yr

(C) 7.5 yr

(D) 10 yr

66. A heat recovery ventilator is placed in a system between exhaust air and make-up air. The ventilator has an effectiveness of 40% and can recover only sensible heat. Air is exhausted at 75°F dry-bulb and 50% relative humidity, and makeup air enters at 35°F dry-bulb and 20% relative humidity. The rates of the exhaust and make-up airstreams are equal. Most nearly, the heat per unit mass that can be recovered is _____.

(A) 3 Btu/lbm

(B) 4 Btu/lbm

(C) 5 Btu/lbm

(D) 6 Btu/lbm

flow in the new duct, the diameter needed is most nearly

(A) 24 in

(B) 40 in

(C) 48 in

(D) 52 in

69. A contractor is preparing to install a fresh air intake on the loading dock as shown.

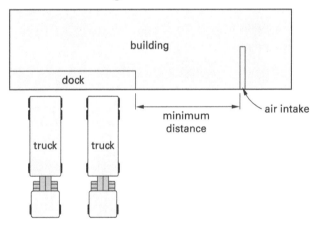

top view

The minimum distance the intake should be placed to prevent truck exhaust from entering the intake is most nearly

(A) 15 ft

(B) 25 ft

(C) 35 ft

(D) 50 ft

70. A round duct industrial ventilation system is shown. The static pressure at the branch is 0.10 in wg, and the exit pressure at each diffuser is 0.03 in wg. The straight branch is 20 ft long with a diameter of 11 in, and the take-off branch is 35 ft long. The flow rate of air needed in each branch is 1000 ft³/min. The friction factor for the ducts is 0.02. Assume standard dry air conditions at sea level, and that the flow rate is not impacted by the 90° turn. Dynamic losses are negligible.

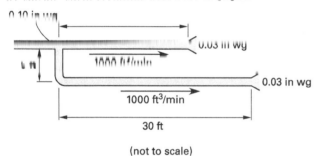

(not to scale)

The duct diameter needed for the take-off branch is most nearly

(A) 8 in

(B) 10 in

(C) 13 in

(D) 18 in

71. Refrigerant-134a (R-134a) flows in a refrigeration system at 10 lbm/min. The refrigerant enters the evaporator at 20 psig (or 34.7 psia) with 15% quality, and it leaves the evaporator at 15 psig (or 29.7 psia) saturated. The refrigerating capacity of the system is most nearly

(A) 4 tons

(B) 5 tons

(C) 6 tons

(D) 7 tons

72. Which of the following air waste heat recovery systems have the potential to recover both latent and sensible heat?

(A) energy recovery ventilator

(B) heat pump

(C) heat recovery ventilator

(D) heat pipe

73. A 40 ton, direct expansion air conditioning system has an average electrical load of 55 kW. The energy efficiency ratio (EER) is most nearly

(A) 8.7

(B) 9.7

(C) 11

(D) 12

74. An ideal, vapor compression refrigeration cycle using refrigerant-134a operates with a 90°F saturation condensing temperature and a 40°F saturation evaporator temperature. There is 10°F subcooling in the condenser and 20°F superheat at the end of the evaporator. The ideal horsepower needed for the compressor to achieve 3 tons of cooling is most nearly

(A) 0.80 hp

(B) 1.3 hp

(C) 2.4 hp

(D) 3.5 hp

75. A local health club with a floor area of 8000 ft² uses air to control the humidity in the building. Air at 95°F and 80% humidity enters an air cooling coil and leaves at 65°F and 20% humidity. The mass flow rate of condensate leaving the airstream is most nearly

(A) 650 lbm/hr

(B) 700 lbm/hr

(C) 750 lbm/hr

(D) 800 lbm/hr

76. A building's dedicated outdoor air system (DOAS) is designed to handle the entire latent load requirement for both the outside air and the building. For winter heating, the supply air condition for the building is 20,000 cfm at 105°F dry-bulb temperature and 25% relative humidity. The inside design condition is 75°F dry-bulb temperature and 45% relative humidity. The outside air requirement is 4000 cfm with winter design conditions of 35°F dry-bulb and 20% relative humidity. The dry-bulb temperature of the supply air for the DOAS is 75°F. The needed relative humidity of the supply air for the DOAS is most nearly

(A) 60%
(B) 70%
(C) 80%
(D) 90%

77. A building's HVAC system includes an air-side economizer. The economizer should be turned on when the

(A) temperature of outside air is lower than the temperature of the supply air
(B) humidity of the outside air is lower than the humidity of the room design condition
(C) enthalpy of the outside air is less than the enthalpy of the supply air
(D) enthalpy of the outside air is less than the enthalpy of the room design condition

78. Hot water at 120°F is pumped through a thin-walled copper pipe that terminates at a boiler. The dimensions of the copper pipe are 2 in × 30 ft long. The water exits the boiler at 280°F through a thin-walled steel pipe that is 3 in × 20 ft long. The copper and steel pipes are exposed to surrounding air at 80°F. The heat loss from the pipes to the surrounding air is most nearly

(A) 112,000 Btu/day
(B) 179,000 Btu/day
(C) 203,000 Btu/day
(D) 266,000 Btu/day

79. A cooling tower with rotational speeds up to 400 rpm is being mounted on a 40 ft × 40 ft concrete pad which is installed on a floor. To control vibration in the cooling tower, which of the following types of isolator is appropriate?

(A) rubber floor
(B) restrained spring
(C) thrust restraint
(D) air spring

80. The ventilation system for a building treats outdoor air of ... air to the building. The rate at which the air is most nearly

(A) 3400 ft³/min
(B) 4100 ft³/min
(C) 6300 ft³/min
(D) 7000 ft³/min

Answer Key

1. C
2. B
3. C
4. B
5. C
6. B
7. D
8. A
9. D
10. B
11. B
12. C
13. B
14. A
15. C
16. B
17. A
18. B
19. A
20. A
21. D
22. A
23. C
24. B
25. C
26. C
27. D
28. C
29. B
30. D
31. D
32. B
33. D
34. A
35. D
36. B
37. C
38. C
39. C
40. B
41. D
42. D
43. C
44. B
45. B
46. B
47. D
48. B
49. C
50. B
51. D
52. C
53. D
54. A
55. B
56. D
57. A
58. C
59. D
60. A
61. C
62. C
63. C
64. C
65. A
66. B
67. C
68. B
69. B
70. C
71. A
72. B, C, D
73. A
74. B
75. D
76. D
77. C
78. D
79. B
80. B

Solutions
Practice Exam

1. The total dynamic head provided by the pump, h_{pump}, is that required to overcome the pressure of the pressurized tank, the friction and fitting losses, h_{losses}, and the static discharge head to the pressurized tank waterline, less the positive static head of the water column to the storage tank above.

Bernoulli's equation describes the energy (as head) relationships as

Bernoulli Equation

$$\frac{p_1}{\gamma} + z_1 + \frac{v_1^2}{2g} = \frac{p_2}{\gamma} + z_2 + \frac{v_2^2}{2g} + h_f$$

In this equation, $h_f = h_{losses} - h_{pump}$, where h_{losses} represents the pipe head loss and h_{pump} represents the pump head.

Rearranging gives

$$h_{pump} + \frac{p_1}{\gamma} + \frac{v_1^2}{2g} + z_1 = \frac{p_2}{\gamma} + \frac{v_2^2}{2g} + z_2 + h_{losses}$$

The equation can be rearranged to solve for the required pump head. The water surface in the open tank is point 1, and the water surface of the pressurized tank is point 2. The velocities v_1 and v_2 are approximately zero. The specific weight of water, γ, is 62.4 lbf/ft^3. [Properties of Water at Standard Conditions]

$$h_{pump} = \frac{p_2 - p_1}{\gamma} + \frac{v_2^2 - v_1^2}{2g} + z_2 - z_1 + h_{losses}$$

$$= \frac{\left(15\ \frac{lbf}{in^2} - 0\ \frac{lbf}{in^2}\right)\left(12\ \frac{in}{ft}\right)^2}{62.4\ \frac{lbf}{ft^3}}$$
$$+ 7\ \text{ft water} - 15\ \text{ft water}$$
$$+ 12\ \text{ft water}$$
$$= 38.62\ \text{ft water} \quad (39\ \text{ft water})$$

The answer is (C).

2. The heat added by the compressor comes from the work done by the compressor. The coefficient of performance (COP) of the process is given as 4.4. The COP is the refrigeration effect divided by the power.

Use the equation for the volumetric flow rate of condensing water and solve for the heat rejection rate, q_o. The density and specific heat capacity of water at standard conditions are 62.4 lbm/ft^3 and 1 Btu/lbf-°F, respectively. [Common Thermodynamic Cycles][Properties of Water at Standard Conditions]

Water-Cooled Condensers

$$Q = \frac{q_o}{\rho C_p (T_2 - T_1)}$$

$$q_o = Q\rho C_p (T_2 - T_1)$$

$$= \left(\frac{\left(640\ \frac{gal}{min}\right)\left(60\ \frac{min}{hr}\right)}{7.481\ \frac{gal}{ft^3}}\right)\left(62.4\ \frac{lbm}{ft^3}\right)$$

$$\times \left(1\ \frac{Btu}{lbm\text{-}°F}\right)(55°F - 43°F)$$

$$= 3{,}844{,}000\ \text{Btu/hr}$$

Use the equation for COP and solve for the input energy in tons using a conversion factor of 12,000 Btu per ton-hr. [Measurement Relationships]

Efficiency

$$\text{COP} = \frac{\text{capacity}}{\text{input energy}}$$

The compressor power, converted to tons of refrigeration, is

$$\text{input energy} = \frac{\text{capacity}}{\text{COP}} = \frac{q_o}{\text{COP}}$$

$$= \frac{3{,}844{,}000\ \frac{Btu}{hr}}{(4.4)\left(12{,}000\ \frac{Btu}{\text{ton-hr}}\right)}$$

$$= 72.8\ \text{tons} \quad (73\ \text{tons})$$

The answer is (B).

3. Convert the units from centistokes to m^2/sec. [Measurement Relationships]

$$(510 \text{ cSt})\left(1 \times 10^{-6} \frac{m^2}{\text{sec} \cdot \text{cSt}}\right) = 5.1 \times 10^{-4} \text{ m}^2/\text{sec}$$

The answer is (C).

4. The reheat load is the rate of sensible heat required to bring the air leaving the coil to the conditions required to supply the gallery room. The density of air at atmospheric pressure and 60°F is 0.0763 lbm/ft³. [Properties of Air at Atmospheric Pressure]

$$q_{\text{reheat}} = \dot{m}c_p(T_{\text{leaving heating coil}} - T_{\text{entering heating coil}})$$

$$= \left(8000 \frac{\text{ft}^3}{\text{min}}\right)\left(60 \frac{\text{min}}{\text{hr}}\right)\left(0.0763 \frac{\text{lbm}}{\text{ft}^3}\right)$$

$$\times \left(0.24 \frac{\text{Btu}}{\text{lbm-°F}}\right)(60°F - 53°F)$$

$$= 61{,}528 \text{ Btu/hr} \quad (61{,}000 \text{ Btu/hr})$$

Alternate Solution

A psychrometric chart can also be used to determine the total energy change of the air, but the sensible heat load might have to be separated out from the total energy change if the process is anything other than pure sensible heating. From the psychrometric chart, at 53°F db and 52°F wb, the total energy content (i.e., the enthalpy) of the air leaving the cooling coil and entering the heating coil is $h_1 = 21.39$ Btu/lbm. The enthalpy of the air leaving the heating coil (and entering the museum) at 60°F db and 51°F dp is $h_2 = 23.02$ Btu/lbm. Points 1 and 2 are nearly on a horizontal line, so the process is essentially pure sensible heating. The specific volume of the air is evaluated at point 2, since 8000 ft³/min is referenced to that point. From the psychrometric chart, $v_2 = 13.26$ ft³/lbm. [ASHRAE Psychrometric Chart No. 1 - Normal Temperature at Sea Level]

The sensible heat load is

$$q_{\text{reheat}} = \dot{m}\Delta h$$

$$= \frac{Q\Delta h}{v}$$

$$= \frac{\left(8000 \frac{\text{ft}^3}{\text{min}}\right)\left(23.02 \frac{\text{Btu}}{\text{lbm}} - 21.39 \frac{\text{Btu}}{\text{lbm}}\right)}{13.26 \frac{\text{ft}^3}{\text{lbm}}} \times \left(60 \frac{\text{min}}{\text{hr}}\right)$$

$$= 59{,}004 \text{ Btu/hr} \quad (60{,}000 \text{ Btu/hr})$$

The answer is (B).

5. For steady flow of an incompressible fluid in a horizontal pipe of constant diameter, the loss of head due to friction is given by the Darcy-Weisbach equation.

Darcy-Weisbach Equation

$$h_f = f\left(\frac{L}{D}\right)\left(\frac{v^2}{2g}\right)$$

From the relationship $\Delta p = \rho g \Delta h$,

$$\Delta h = \frac{\Delta p}{\rho g}$$

Substituting into the Darcy-Weisbach equation gives

$$\frac{\Delta p}{\rho g} = f\left(\frac{L}{D}\right)\left(\frac{v^2}{2g}\right)$$

$$\Delta p = \frac{fLv^2\rho}{2D}$$

When flow is laminar, friction factor is a function of the Reynolds number.

$$f = \frac{64}{\text{Re}} = \frac{64\mu}{vD\rho}$$

Velocity equals

$$v = \frac{Q}{A} = \frac{4Q}{\pi D^2}$$

Substituting into the equation for friction factor gives

$$f = \frac{64\mu}{\left(\frac{4Q}{\pi D^2}\right)D\rho} = \frac{16\pi\mu D}{Q\rho}$$

Substituting this into the equation for pressure drop gives

$$\Delta p = \frac{\left(\frac{16\pi\mu D}{Q\rho}\right)L\mathrm{v}^2\rho}{2D} = \frac{8\pi\mu L\mathrm{v}^2}{Q}$$

$$= \frac{8\pi\mu L\left(\frac{Q}{A}\right)^2}{Q}$$

$$= \frac{8\pi\mu L\left(\frac{4Q}{\pi D^2}\right)^2}{Q}$$

$$= \frac{128\mu L Q}{\pi D^4}$$

This equation for pressure drop is the Hagen-Poiseuille equation, also called Poiseuille's law.

$$\Delta p = \frac{(128)\left(26.37\times 10^{-6}\,\frac{\text{lbf-sec}}{\text{ft}^2}\right)}{\pi (0.018\text{ in})^4}$$
$$\times\left(3\times 10^{-5}\,\frac{\text{ft}^3}{\text{sec}}\right)(3.4\text{ ft})\left(12\,\frac{\text{in}}{\text{ft}}\right)^4$$
$$= 21{,}647\text{ lbf/ft}^2 \quad (22{,}000\text{ lbf/ft}^2)$$

The answer is (C).

6. The interest rate is unknown. Using the straight-line depreciation method,

Depreciation: Straight Line

$$D_j = \frac{C - S_n}{n} = \frac{\$525{,}000 - \$0}{15} = \$35{,}000$$

C is the purchase price or cost, S_n is the expected salvage value, and n is the number of years. The book value, BV, can be calculated at 7 years.

$$\text{BV}_7 = C - jD_j = \$525{,}000 - (7)(\$35{,}000) = \$280{,}000$$

The gain on the sale of a depreciated asset is

$$\text{gain} = S - \text{BV}_7 = \$700{,}000 - \$280{,}000 = \$420{,}000$$

So,

$$\begin{aligned}P = &-\$525{,}000 + (\$700{,}000)(P/F, i\%, 7)\\ &+(\$35{,}000)(P/A, i\%, 7)(0.40)\\ &-(\$420{,}000)(P/F, i\%, 7)(0.40)\\ &-(\$25{,}000)(P/A, i\%, 7)(1-0.40)\\ &+(\$45{,}000)(P/A, i\%, 3)(1-0.40)\\ &+(\$75{,}000)(P/A, i\%, 7)(1-0.40)\\ &-(\$75{,}000)(P/A, i\%, 3)(1-0.40)\end{aligned}$$

Neglect the benefit of a favorable capital gains tax rate on the gain above $525,000.

Simplify and combine terms.

$$\begin{aligned}P = &-\$525{,}000\\ &+\begin{pmatrix}\$700{,}000\\ -(\$420{,}000)(0.40)\end{pmatrix}(P/F, i\%, 7)\\ &+\begin{pmatrix}(\$35{,}000)(0.40)\\ -(\$25{,}000)(0.60)\\ +(\$75{,}000)(0.60)\end{pmatrix}(P/A, i\%, 7)\\ &+\begin{pmatrix}(\$45{,}000)(0.60)\\ -(\$75{,}000)(0.60)\end{pmatrix}(P/A, i\%, 3)\\ = &-\$525{,}000 + (\$532{,}000)(P/F, i\%, 7)\\ &+(\$44{,}000)(P/A, i\%, 7)\\ &+(-\$18{,}000)(P/A, i\%, 3)\end{aligned}$$

Try $i = 10\%$. For a period of 7 years, the economic cost factor is 0.5132. [Economic Factor Tables]

$$\begin{aligned}P = &-\$525{,}000 + (\$532{,}000)(0.5132)\\ &+(\$44{,}000)(4.8684)\\ &+(-\$18{,}000)(2.4869)\\ = &-\$82{,}532\end{aligned}$$

Try $i = 6\%$. For a period of 7 years, the economic cost factor is 0.6651. [Economic Factor Tables]

$$\begin{aligned}P = &-\$525{,}000 + (\$532{,}000)(0.6651)\\ &+(\$44{,}000)(5.5824)\\ &+(-\$18{,}000)(2.6730)\\ = &\$26{,}345\end{aligned}$$

Use linear interpolation. The after-tax rate of return is

$$\text{ROR} \approx 6\% + (10\% - 6\%)$$
$$\times \left(\frac{\$26{,}345}{\$26{,}345 - (-\$82{,}532)} \right)$$
$$= 6.96\% \quad (7.0\%)$$

The answer is (B).

7. The Bernoulli equation can be used to solve this problem.

Bernoulli Equation

$$\frac{p_1}{\gamma} + z_1 + \frac{v_1^2}{2g} = \frac{p_2}{\gamma} + z_2 + \frac{v_2^2}{2g} + h_f$$

In this equation, $h_f = h_{\text{losses}} - h_{\text{pump}}$, where h_{losses} represents the pipe head loss and h_{pump} represents the pump head. Rearranging gives

$$h_{\text{pump}} = \frac{p_2 - p_1}{\gamma} + \frac{v_2^2 - v_1^2}{2g} + z_2 - z_1 + h_{\text{losses}}$$

The flow rate in each line is 7000 gpm.

$$Q = \frac{7000 \ \frac{\text{gal}}{\text{min}}}{\left(7.481 \ \frac{\text{gal}}{\text{ft}^3} \right) \left(60 \ \frac{\text{sec}}{\text{min}} \right)}$$
$$= 15.60 \ \text{ft}^3/\text{sec}$$

Use the continuity equation to find the velocities in the suction and discharge lines.

Continuity Equation

$$Q = A v$$
$$v = \frac{Q}{A} = \frac{4Q}{\pi D^2}$$

For the suction line,

$$v_1 = \frac{4Q}{\pi D_1^2}$$
$$= \frac{(4) \left(15.60 \ \frac{\text{ft}^3}{\text{sec}} \right)}{\pi \left(\frac{20 \ \text{in}}{12 \ \frac{\text{in}}{\text{ft}}} \right)^2}$$
$$= 7.151 \ \text{ft/sec}$$

For the discharge line,

$$v_2 = \frac{4Q}{\pi D_2^2}$$
$$= \frac{(4) \left(15.60 \ \frac{\text{ft}^3}{\text{sec}} \right)}{\pi \left(\frac{12 \ \text{in}}{12 \ \frac{\text{in}}{\text{ft}}} \right)^2}$$
$$= 19.86 \ \text{ft/sec}$$

Solving the rearranged Bernoulli equation gives

$$h_{\text{pump}} = \frac{p_2 - p_1}{\gamma} + \frac{v_2^2 - v_1^2}{2g} + z_2 - z_1 + h_{\text{losses}}$$
$$= \frac{\left(500 \ \frac{\text{lbf}}{\text{in}^2} - 100 \ \frac{\text{lbf}}{\text{in}^2} \right) \left(12 \ \frac{\text{in}}{\text{ft}} \right)^2}{62.4 \ \frac{\text{lbf}}{\text{ft}^3}}$$
$$+ \frac{19.86 \ \frac{\text{ft}}{\text{sec}} - 7.151 \ \frac{\text{ft}}{\text{sec}}}{(2) \left(32.2 \ \frac{\text{ft}}{\text{sec}^2} \right)}$$
$$+ 30 \ \text{ft} - 0 \ \text{ft} + 10 \ \text{ft}$$
$$= 963.3 \ \text{ft}$$

The power needed to produce this amount of head is

Centrifugal Pump Characteristics

$$P_{\text{fluid}} = \rho g h Q$$
$$= \gamma h_{\text{pump}} Q$$
$$= \frac{\left(62.4 \ \frac{\text{lbf}}{\text{ft}^3} \right) (963.3 \ \text{ft}) \left(\frac{7000 \ \frac{\text{gal}}{\text{min}}}{7.481 \ \frac{\text{gal}}{\text{ft}^3}} \right)}{33{,}000 \ \frac{\text{ft-lbf}}{\text{hp-min}}}$$
$$= 1704 \ \text{hp} \quad (1700 \ \text{hp})$$

The answer is (D).

8. The total resistance of the window glass includes the glass R-value plus the indoor and outdoor surface resistances. [Composite Plane Wall]

$$R_{\text{total}} = R_{\text{outdoor air film}} + R_{\text{glass}} + R_{\text{indoor air film}}$$

The temperature drop between indoor and outdoor temperatures at a point in an assembly is proportional to the R-value at that point relative to the total R-value.

Composite Plane Wall

$$\frac{R_x}{R_{\text{total}}} = \frac{(T_{\text{inside}} - T_x)}{T_{\text{inside}} - T_{\text{outside}}}$$

$$\Delta T_x = \frac{\Delta T_{\text{total}} R_x}{R_{\text{total}}}$$

The total resistance of the glass is the sum of the resistance of the glass and the reciprocal of the indoor and outdoor surface conductances.

$$R_{\text{total}} = \frac{1}{6.0 \frac{\text{Btu}}{\text{hr-ft}^2\text{-}°\text{F}}} + 1.5 \frac{\text{hr-ft}^2\text{-}°\text{F}}{\text{Btu}}$$

$$+ \frac{1}{1.46 \frac{\text{Btu}}{\text{hr-ft}^2\text{-}°\text{F}}}$$

$$= 2.35 \text{ hr-ft}^2\text{-}°\text{F/Btu}$$

The indoor surface temperature of the glass is calculated by subtracting the temperature drop due to the resistance of the indoor air film from room temperature.

$$\Delta T_{\text{indoor air film}} = (70°\text{F} - (-40°\text{F})) \left(\frac{\frac{1}{1.46 \frac{\text{Btu}}{\text{hr-ft}^2\text{-}°\text{F}}}}{2.35 \frac{\text{hr-ft}^2\text{-}°\text{F}}{\text{Btu}}} \right)$$

$$= 32.06°\text{F}$$

$$T_{\text{surface}} = T_{\text{room}} - \Delta T_{\text{indoor air film}}$$
$$= 70°\text{F} - 32.06°\text{F}$$
$$= 37.94°\text{F} \quad (38°\text{F})$$

To avoid condensation, the room air dew point must be less than 38°F. The corresponding relative humidity at room temperature is determined by locating 38°F on the saturation curve of the psychrometric chart and following a line of constant humidity to the right to 70°F. [ASHRAE Psychrometric Chart No. 1 - Normal Temperature at Sea Level]

This state point corresponds to approximately 31% relative humidity.

The answer is (A).

9. The absolute temperature at the beginning of the process is

$$T_1 = 100°\text{F} + 460° = 560°\text{R}$$

Polytropic processes follow the same relationships as isentropic processes, but using a polytropic coefficient, n, in place of the ratio of specific heats, k. For an isentropic process,

Isentropic Flow Relationships

$$\frac{p_2}{p_1} = \left(\frac{T_2}{T_1}\right)^{k/(k-1)}$$

For a polytropic process,

$$\frac{p_2}{p_1} = \left(\frac{T_2}{T_1}\right)^{n/(n-1)}$$

$$T_2 = T_1 \left(\frac{p_2}{p_1}\right)^{(n-1)/n}$$

$$= (560°\text{R}) \left(\frac{40 \frac{\text{lbf}}{\text{in}^2}}{120 \frac{\text{lbf}}{\text{in}^2}}\right)^{(1.6-1)/1.6}$$

$$= 370.9°\text{R}$$

The work done by the system per pound-mole is given by

Special Cases of Closed Systems (With No Change in Kinetic or Potential Energy)

$$W = \frac{(p_2 V_2 - p_1 V_1)}{1 - n}$$

Using knowledge of ideal gases,

$$W = \frac{R(T_1 - T_2)}{n-1} = \frac{\left(1545 \frac{\text{ft-lbf}}{\text{lbmol-}°\text{R}}\right)(560°\text{R} - 370.9°\text{R})}{1.6 - 1}$$

$$= 4.87 \times 10^5 \text{ ft-lbf/lbmol}$$

The molecular weight of air is 29 lbm/lbmol. [Thermal and Physical Properties of Ideal Gases (at Room Temperature)]

The work per unit mass is

$$W = \frac{4.87 \times 10^5 \frac{\text{ft-lbf}}{\text{lbmol}}}{29 \frac{\text{lbm}}{\text{lbmol}}}$$

$$= 1.679 \times 10^4 \text{ ft-lbf/lbm} \quad (1.7 \times 10^4 \text{ ft-lbf/lbm})$$

The answer is (D).

10. Convert the resistivity to square inches. A circular mil is the area of a circle with a diameter of 0.001 in.

$$A_{\text{per cmil}} = \left(\frac{\pi}{4}\right)D^2 = \left(\frac{\pi}{4}\right)(0.001 \text{ in})^2$$
$$= 7.854 \times 10^{-7} \text{ in}^2$$

$$\rho = R_{\text{electrical}} A$$
$$= \left(10 \, \frac{\Omega\text{-cmil}}{\text{ft}}\right)\left(7.854 \times 10^{-7} \, \frac{\text{in}^2}{\text{cmil}}\right)$$
$$= 7.854 \times 10^{-6} \, \Omega\text{-in}^2/\text{ft}$$

The cross-sectional area of the copper tubing is

$$A_{\text{tube}} = \left(\frac{\pi}{4}\right)(D_o^2 - D_i^2)$$
$$= \left(\frac{\pi}{4}\right)((3.5 \text{ in})^2 - (3.062 \text{ in})^2)$$
$$= 2.257 \text{ in}^2$$

The maximum resistance that will lead to a drop of no more than 0.6 V is

Ohm's Law
$$V = IR$$
$$R_{\text{max}} = \frac{V_{\text{max}}}{I} = \frac{0.6 \text{ V}}{35 \text{ A}} = 0.01714 \, \Omega$$

The resistivity of a material is

Electrical Properties
$$\rho = \frac{RA}{L}$$

The maximum length of the circuit is

$$L_{\text{max}} = \frac{R_{\text{max}} A_{\text{tube}}}{\rho}$$
$$= \frac{(0.01714 \, \Omega)(2.257 \text{ in}^2)}{7.854 \times 10^{-6} \, \frac{\Omega\text{-in}^2}{\text{ft}}}$$
$$= 4926 \text{ ft}$$

Two wires (pipes) are needed to complete the circuit, so the maximum distance between power supply and pump is

$$D_{\text{max}} = \frac{L_{\text{max}}}{2} = \frac{4926 \text{ ft}}{2} = 2463 \text{ ft} \quad (2500 \text{ ft})$$

The answer is (B).

11. Find the change in temperature. For a single-phase pure component, specifying any two intensive, independent properties is sufficient to determine all the rest. The heat capacity at constant volume is

$$c_v = \left(\frac{\partial u}{\partial T}\right)_v = \frac{\Delta u}{\Delta T}$$
$$\Delta u = c_v \Delta T$$

Multiply both sides by the mass to find the total change in internal energy, then rearrange to solve for the change in temperature.

$$\Delta U = m c_v \Delta T$$
$$\Delta T = \frac{\Delta U}{m c_v}$$

To find the change in internal energy, ΔU, use the first law of thermodynamics.

Closed Thermodynamic Systems
$$Q - W = \Delta U + \Delta KE + \Delta PE$$

To achieve the maximum possible change in temperature, assume the tank is perfectly insulated so that $Q = 0$, as does ΔKE and ΔPE. W is negative because work is being done on the system.

$$-W = \Delta U$$

Find the total work done by the mixing motor. [Measurement Relationships]

$$W = Pt$$
$$= (2 \text{ hp})\left(550 \, \frac{\text{ft-lbf}}{\text{hp-sec}}\right)(15 \text{ min})\left(60 \, \frac{\text{sec}}{\text{min}}\right)$$
$$= 990{,}000 \text{ ft-lbf}$$

Convert ft-lbf to Btu. [Measurement Relationships]

$$\Delta U = \frac{990{,}000 \text{ ft-lbf}}{778 \, \frac{\text{ft-lbf}}{\text{Btu}}}$$
$$= 1272 \text{ Btu}$$

Find the maximum possible rise in temperature. The specific heat of water is 1.0 Btu/lbm-°F. [Properties of Water at Standard Conditions]

$$\Delta T = \frac{\Delta U}{m c_v} = \frac{1272 \text{ Btu}}{(60 \text{ lbm})\left(1.0 \, \frac{\text{Btu}}{\text{lbm-°F}}\right)}$$
$$= 21.2°\text{F} \quad (21°\text{F})$$

The answer is (B).

12. Draw a diagram and a thermal circuit of the system.

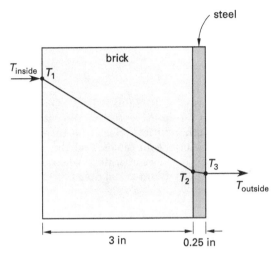

Use the given convection coefficient of $\bar{h}_{co} = 1.65$ Btu/hr-ft²-°F. The thermal resistance, R_{co}, per square foot is

Thermal Resistance (R)

$$R_{co} = \frac{1}{\bar{h}_{co}A} = \frac{1}{\left(1.65 \, \frac{\text{Btu}}{\text{hr-ft}^2\text{-°F}}\right)(1 \text{ ft}^2)}$$

$$= 0.606 \text{ hr-°F/Btu}$$

$$k_{br} = 0.58 \text{ Btu-ft/hr-ft}^2\text{-°F}$$

$$k_s = 26 \text{ Btu-ft/hr-ft}^2\text{-°F}$$

(The value of k_s depends on the steel composition, but it does not affect the final result.)

Thermal Resistance (R)

$$R_{br} = \frac{L_{br}}{k_{br}A} = \frac{3 \text{ in}}{\left(0.58 \, \frac{\text{Btu-ft}}{\text{hr-ft}^2\text{-°F}}\right)(1 \text{ ft}^2)\left(12 \, \frac{\text{in}}{\text{ft}}\right)}$$

$$= 0.431 \text{ hr-°F/Btu}$$

$$R_s = \frac{L_s}{k_sA} = \frac{0.25 \text{ in}}{\left(26 \, \frac{\text{Btu-ft}}{\text{hr-ft}^2\text{-°F}}\right)(1 \text{ ft}^2)\left(12 \, \frac{\text{in}}{\text{ft}}\right)}$$

$$= 0.0008 \text{ hr-°F/Btu}$$

$$T_1 = 1000\text{°F}$$

$$T_\infty = 70\text{°F}$$

Since T_1 and T_∞ are the only known temperatures, the heat transfer is

Composite Plane Wall

$$q = UA(T_{\text{inside}} - T_{\text{outside}})$$

$$U = \frac{1}{R_{\text{total}}}$$

Solving for the heat transfer rate gives

$$q = \frac{T_1 - T_\infty}{R_{br} + R_s + R_{co}}$$

$$= \frac{1000\text{°F} - 70\text{°F}}{0.431 \, \frac{\text{hr-°F}}{\text{Btu}} + 0.0008 \, \frac{\text{hr-°F}}{\text{Btu}} + 0.606 \, \frac{\text{hr-°F}}{\text{Btu}}}$$

$$= 896 \text{ Btu/hr}$$

The outside surface temperature of the steel is

$$q = \frac{T_3 - T_\infty}{R_{co}}$$

$$T_3 = qR_{co} + T_{\text{inf}}$$

$$= \left(896 \, \frac{\text{Btu}}{\text{hr}}\right)\left(0.606 \, \frac{\text{hr-°F}}{\text{Btu}}\right) + 70\text{°F}$$

$$= 613\text{°F} \quad (610\text{°F})$$

The answer is (C).

13. The temperature distribution in a single-pass, parallel flow heat exchanger is shown.

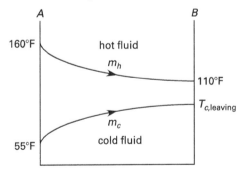

Find the temperature of the cooling fluid leaving the heat exchanger.

$$q_h = \dot{m}_h c_{p,h} \Delta T_h$$
$$= \left(45{,}000 \; \frac{\text{lbm}}{\text{hr}}\right)\left(0.9 \; \frac{\text{Btu}}{\text{lbm-°F}}\right)(160\,°\text{F} - 110\,°\text{F})$$
$$= 2.025 \times 10^6 \; \frac{\text{Btu}}{\text{hr}}$$

$$q_c = \dot{m}_c c_{p,c} \Delta T_c$$
$$= \left(40{,}000 \; \frac{\text{lbm}}{\text{hr}}\right)\left(1 \; \frac{\text{Btu}}{\text{lbm-°F}}\right)(T_{c,\text{leaving}} - 55\,°\text{F})$$
$$= \left(40{,}000 \; \frac{\text{Btu}}{\text{hr-°F}}\right)(T_{c,\text{leaving}} - 55\,°\text{F})$$

The two heat flow rates are equal. Therefore,

$$q_h = q_c$$
$$2.025 \times 10^6 \; \frac{\text{Btu}}{\text{hr}} = \left(40{,}000 \; \frac{\text{Btu}}{\text{hr-°F}}\right)(T_{c,\text{leaving}} - 55\,°\text{F})$$
$$T_{c,\text{leaving}} = 105.6\,°\text{F}$$

The log mean temperature difference, ΔT_{lm}, is

Log Mean Temperature Difference (LMTD)

$$\Delta T_{\text{lm}} = \frac{(T_{Ho} - T_{Co}) - (T_{Hi} - T_{Ci})}{2.3 \log_{10} \dfrac{T_{Ho} - T_{Co}}{T_{Hi} - T_{Ci}}}$$

$$= \frac{(110\,°\text{F} - 105.6\,°\text{F}) - (160\,°\text{F} - 55\,°\text{F})}{2.3 \log_{10} \dfrac{110\,°\text{F} - 105.6\,°\text{F}}{160\,°\text{F} - 55\,°\text{F}}}$$

$$= 31.7\,°\text{F}$$

Use the equation for the rate of heat transfer in a heat exchanger, and solve for the heat transfer surface area. The correction factor, F, is 1.

Rate of Heat Transfer

$$\dot{Q} = UAF\Delta T_{\text{lm}}$$
$$A = \frac{\dot{Q}}{UF\Delta T_{\text{lm}}}$$

$$= \frac{2.025 \times 10^6 \; \dfrac{\text{Btu}}{\text{hr}}}{\left(75 \; \dfrac{\text{Btu}}{\text{hr-ft}^2\text{-°F}}\right)(1)(31.7\,°\text{F})}$$

$$= 851.7 \; \text{ft}^2 \quad (850 \; \text{ft}^2)$$

The answer is (B).

14. Define states 1 and 2' as shown.

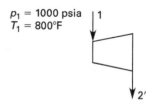

Using steam tables, $h_1 = 1389.0$ Btu/lbm, $s_1 = 1.567$ Btu/lbm-°R, and $p_2 = 4$ psia. h_2 represents the enthalpy for a turbine that is 100% efficient. Since the turbine is isentropic, $s_1 = s_2$. Using steam tables, find the appropriate enthalpy and entropy values at state 2' where 2' = 4 psia. [Properties of Saturated Water and Steam (Temperature) - I-P Units]

$$h_f = 120.87 \; \text{Btu/lbm}$$
$$s_f = 0.2198 \; \text{Btu/lbm-°R}$$
$$h_{fg} = 1006.4 \; \text{Btu/lbm}$$
$$s_{fg} = 1.6424 \; \text{Btu/lbm-°R}$$

The steam quality at the turbine exhaust (state 2) for a 100% efficient turbine is found from the entropy relationship.

Properties for Two-Phase (Vapor-Liquid) Systems

$$s = s_f + x s_{fg}$$
$$x = \frac{s - s_f}{s_{fg}}$$

$$= \frac{1.567 \; \dfrac{\text{Btu}}{\text{lbm-°R}} - 0.2198 \; \dfrac{\text{Btu}}{\text{lbm-°R}}}{1.6424 \; \dfrac{\text{Btu}}{\text{lbm-°R}}}$$

$$= 0.82$$

Use the steam quality to find the enthalpy at state 2, h_2.

Properties for Two-Phase (Vapor-Liquid) Systems

$$h_2 = h_f + x h_{fg} = 120.87 \; \frac{\text{Btu}}{\text{lbm}} + (0.82)\left(1006.4 \; \frac{\text{Btu}}{\text{lbm}}\right)$$
$$= 945.8 \; \text{Btu/lbm}$$

Since the turbine exhaust steam quality is 100%, the enthalpy at state 2' is equal to the enthalpy of saturated vapor, h_g. From the steam tables at 4 psia,

$$h_2' = h_g = 1126.91 \; \text{Btu/lbm}$$

The thermal efficiency of the turbine is

Special Cases of the Steady-Flow Energy Equation

$$\eta_s = \frac{h_i - h_e}{h_i - h_{es}} = \frac{h_1 - h_{2'}}{h_1 - h_2}$$

$$= \frac{1389.0 \frac{\text{Btu}}{\text{lbm}} - 1126.91 \frac{\text{Btu}}{\text{lbm}}}{1389.0 \frac{\text{Btu}}{\text{lbm}} - 945.8 \frac{\text{Btu}}{\text{lbm}}}$$

$$= 0.59 \quad (59\%)$$

The answer is (A).

15. Write the general combustion reaction for a hydrocarbon, C_nH_m, burning with theoretical air. [Combustion Processes]

$$C_nH_m + xO_2 + 3.76xN_2$$
$$\rightarrow aCO_2 + bH_2O + 3.76xN_2$$

$a = n$ [from carbon balance]
$2b = m$ [from hydrogen balance]
$x = n + \dfrac{m}{4}$ [from oxygen balance]

The composition of methane is CH_4. Therefore, $n=1$, and $m=4$, so $a=1$, $b=2$, and $x=2$. The combustion reaction for the combustion of methane in theoretical air is given by

$$CH_4 + 2O_2 + (3.76)(2)N_2$$
$$\rightarrow CO_2 + 2H_2O + (3.76)(2)N_2$$

For 125% theoretical air, the reaction is

$$CH_4 + (1.25)(2)O_2 + (1.25)(7.52)N_2$$
$$\rightarrow CO_2 + 2H_2O + (0.25)(2)O_2 + (1.25)(7.52)N_2$$

The masses of air and fuel required for the reaction are found using the approximate molecular weights. [Thermal and Physical Properties of Ideal Gases (at Room Temperature)]

$$M_f = (1)\left(16 \frac{\text{g}}{\text{mol}}\right) = 16 \text{ g/mol}$$

$$M_a = ((1.25)(2) + (1.25)(7.52))\left(28.96 \frac{\text{g}}{\text{mol}}\right)$$

$$= 344.6 \text{ g/mol}$$

The air-to-fuel ratio is

$$R_{\text{af}} = \frac{M_a}{M_f} = \frac{344.6 \frac{\text{g}}{\text{mol}}}{16 \frac{\text{g}}{\text{mol}}} = 21.5:1 \quad (22:1)$$

The answer is (C).

16. Since the problem involves transient heat flow, determine if the lumped capacitance approximation is valid. The average temperature of the copper is

$$\overline{T} = \frac{T_s + T_l}{2} = \frac{852°F + 392°F}{2}$$
$$= 622°F$$

The sphere's specific heat and density are found in a table of properties of metals. [Properties of Metals - I-P Units]

$$c_p = 0.09 \text{ Btu/lbm-°F}$$
$$\rho = 557 \text{ lbm/ft}^3$$

Determine the Biot number.

Transient Conduction Using the Lumped Capacitance Model

$$\text{Bi} = \frac{hV}{kA}$$

$$= \frac{h\left(\frac{4}{3}\pi r^3\right)}{k(4\pi r^2)}$$

$$= \frac{hr}{3k}$$

$$= \frac{\left(155 \frac{\text{Btu}}{\text{hr-ft}^2\text{-°F}}\right)\left(\frac{7.88 \text{ in}}{2}\right)}{(3)\left(2628 \frac{\text{Btu-in}}{\text{hr-ft}^2\text{-°F}}\right)\left(12 \frac{\text{in}}{\text{ft}}\right)}$$

$$= 0.0065$$

Because the Biot number is much less than 0.1, the internal thermal resistance of the sphere is negligible compared to the external thermal resistance in the oil bath. Therefore, the lumped parameter method can be used to solve for the time constant.

Constant Fluid Temperature
$$T_t = T_\infty + (T_0 - T_\infty)e^{-\beta t}$$
$$e^{-\beta t} = \frac{T_t - T_\infty}{T_0 - T_\infty}$$
$$-\beta t = \ln\frac{T_t - T_\infty}{T_0 - T_\infty}$$
$$t = -\beta \ln\frac{T_t - T_\infty}{T_0 - T_\infty}$$

The exponent β is

$$\beta = \frac{hA_s}{\rho V c_p}$$
$$= \frac{h(4\pi r^2)}{\rho\left(\frac{4}{3}\pi r^3\right)c_p}$$
$$= \frac{3h}{\rho r c_p}$$
$$= \frac{(3)\left(155\ \frac{\text{Btu}}{\text{hr-ft}^2\text{-°F}}\right)\left(12\ \frac{\text{in}}{\text{ft}}\right)}{\left(557\ \frac{\text{lbm}}{\text{ft}^3}\right)\left(\frac{7.88\ \text{in}}{2}\right)\left(0.09\ \frac{\text{Btu}}{\text{lbm-°F}}\right)\left(3600\ \frac{\text{sec}}{\text{hr}}\right)}$$
$$= 0.00785\ \text{sec}^{-1}$$

Solve for t.

$$t = -\frac{1}{\beta}\ln\frac{T - T_\infty}{T_i - T_\infty}$$
$$= -\left(\frac{1}{0.00785\ \frac{1}{\text{sec}}}\right)\ln\left(\frac{392°\text{F} - 167°\text{F}}{852°\text{F} - 167°\text{F}}\right)$$
$$= 141.82\ \text{sec}\quad (140\ \text{sec})$$

The answer is (B).

17. The useful refrigeration provided by the chiller is determined from the change in enthalpy across the evaporator.

Special Cases of the Steady-Flow Energy Equation
$$h_1 + q = h_e$$

$$\dot{Q}_{\text{refrigeration}} = \dot{m}_{\text{refrigerant}}(h_{\text{leaving}} - h_{\text{entering}})$$

The enthalpy of the refrigerant entering the evaporator on the low-pressure side is equal to the enthalpy of the refrigerant leaving the condenser as a saturated liquid. From refrigerant property tables for R-134a, the enthalpy of saturated liquid at 100 psia is 37.8 Btu/lbm (38 Btu/lbm). [**Refrigerant 134a (1,1,1,2-Tetrafluoroethane) Properties of Saturated Liquid and Saturated Vapor**]

The enthalpy of the refrigerant entering the compressor is determined by locating the state point for saturated vapor at the low-pressure side, extended horizontally 20°F for superheat. The refrigerant temperature at a saturated vapor pressure of 33 psia is 19.8°F (20°F). Extending a horizontal line from the saturated vapor state point at 33 psia to approximately 40°F (midway between the 20°F and 60°F temperature curves) locates the superheat process line. The extension terminates at 110.2 Btu/lbm (110 Btu/lbm). [**Pressure Versus Enthalpy Curves for Refrigerant 134a**]

The mass flow rate is

$$\dot{m}_{\text{refrigerant}} = \frac{\dot{Q}_{\text{refrigeration}}}{h_{\text{leaving}} - h_{\text{entering}}}$$
$$= \frac{(255\ \text{tons})\left(12{,}000\ \frac{\text{Btu}}{\text{hr-ton}}\right)}{110\ \frac{\text{Btu}}{\text{lbm}} - 38\ \frac{\text{Btu}}{\text{lbm}}}$$
$$= 42{,}500\ \text{lbm/hr}\quad (43{,}000\ \text{lbm/hr})$$

The answer is (A).

18. The humidity ratio for the air mixture can be determined graphically on the psychrometric chart using the lever rule or mathematically using the ratio of outside air to mixed air.

From a psychrometric chart, the humidity ratio for outside air at 90°F dry-bulb and 75°F wet-bulb is 0.0153 lbm moisture/lbm dry air. The humidity ratio for air returning from the room at 75°F dry-bulb and 50% relative humidity is 0.0093 lbm moisture/lbm dry air. [**ASHRAE Psychrometric Chart No. 1 - Normal Temperature at Sea Level**]

The total humidity ratio of the air entering the coil is

$$w_{\text{mixed}} = w_{\text{return}} + \left(\frac{Q_{\text{outside}}}{Q_{\text{return}} + Q_{\text{outside}}}\right)(w_{\text{outside}} - w_{\text{return}})$$
$$= 0.0093\ \frac{\text{lbm moisture}}{\text{lbm dry air}}$$
$$+ \left(\frac{2300\ \frac{\text{ft}^3}{\text{min}}}{7000\ \frac{\text{ft}^3}{\text{min}} + 2300\ \frac{\text{ft}^3}{\text{min}}}\right)$$
$$\times \left(\begin{array}{c}0.0153\ \dfrac{\text{lbm moisture}}{\text{lbm dry air}} \\ -0.0093\ \dfrac{\text{lbm moisture}}{\text{lbm dry air}}\end{array}\right)$$
$$= 0.0108\ \text{lbm moisture/lbm dry air}$$
$$(0.011\ \text{lbm moisture/lbm dry air})$$

The answer is (B).

19. The net positive suction head available, NPSHA, for the pump is the positive head available for introducing liquid into the pump without lowering the pressure in the pump below the vapor pressure and causing cavitation. It is determined by subtracting any reductions in head from local atmospheric pressure and subtracting the friction head and vapor pressure head of the water.

$$\text{NPSHA} = h_{\text{atmospheric}} \pm h_{\text{static}} - h_{\text{vaporpressure}} - h_{\text{friction}}$$

Atmospheric pressure is known at sea level, and static and friction head are given. From steam tables, the vapor pressure of the water at 80°F is 0.50683 lbf/in². [Properties of Water at Standard Conditions]

The pressure head is

Duct Design: Bernoulli Equation

$$h = \frac{pg_c}{\rho g} = \frac{p}{\gamma}$$

For the pump,

$$\text{NPSHA} = h_{\text{atmospheric}} \pm h_{\text{static}} - h_{\text{vp}} - h_{\text{friction}}$$

$$= \left(14.7 \frac{\text{lbf}}{\text{in}^2}\right) \left(\frac{\left(12 \frac{\text{in}}{\text{ft}}\right)^2}{62.4 \frac{\text{lbf}}{\text{ft}^3}}\right) - 17 \text{ ft}$$

$$- \left(0.50683 \frac{\text{lbf}}{\text{in}^2}\right) \left(\frac{\left(12 \frac{\text{in}}{\text{ft}}\right)^2}{62.4 \frac{\text{lbf}}{\text{ft}^3}}\right) - 3.0 \text{ ft}$$

$$= 12.8 \text{ ft} \quad (13 \text{ ft water})$$

The answer is (A).

20. Find the initial horsepower.

Pump Power Equation

$$P = Q\rho \left(\frac{g}{g_c}\right) h \left(\frac{1}{\eta_{\text{pump}}}\right)$$

$$= \frac{\left(250 \frac{\text{gal}}{\text{min}}\right)\left(8.34 \frac{\text{lbm}}{\text{gal}}\right)\left(\frac{32.2 \frac{\text{ft}}{\text{sec}^2}}{32.2 \frac{\text{ft-lbm}}{\text{lbf-sec}^2}}\right)(75 \text{ ft})\left(\frac{1}{0.65}\right)}{33{,}000 \frac{\text{ft-lbf}}{\text{hp-min}}}$$

$$= 7.29 \text{ hp}$$

Combine two affinity laws to get an equation for the flow rate when the pump is making full use of the 10 hp motor.

Pump Affinity Laws

$$\text{bhp}_2 = \text{bhp}_1 \left(\frac{N_2}{N_1}\right)^3$$

$$Q_2 = Q_1 \left(\frac{N_2}{N_1}\right)$$

The maximum flow rate is

$$Q_2 = Q_1 \sqrt[3]{\frac{\text{bhp}_2}{\text{bhp}_1}}$$

$$= \left(250 \frac{\text{gal}}{\text{min}}\right) \sqrt[3]{\frac{10 \text{ hp}}{7.29 \text{ hp}}}$$

$$= 277.8 \text{ gpm} \quad (280 \text{ gpm})$$

The answer is (A).

21. The casino has no walls or windows on the building exterior, and the facility runs 24 hr/day. This implies that there will not be significant variation in the cooling load. The lighting load for the space is

Electric Lighting

$$q_{\text{lighting}} = \left(3.412 \frac{\text{Btu}}{\text{hr-W}}\right) W F_{\text{ul}} F_{\text{sa}}$$

$$= \left(3.412 \frac{\text{Btu}}{\text{hr-W}}\right)\left(3.75 \frac{\text{W}}{\text{ft}^2}\right)(40{,}000 \text{ ft}^2)(1)(1)$$

$$= 511{,}950 \text{ Btu/hr}$$

The equipment load for the space is

$$q_{\text{equipment}} = (80 \text{ kW})\left(3412 \frac{\text{Btu}}{\text{hr-kW}}\right) = 273{,}040 \text{ Btu/hr}$$

From an *ASHRAE Handbook* table for heat and moisture rates given off by people in different states of activity, the activities of standing, light work, or walking contribute per capita sensible and latent loads of 250 Btu/hr and 200 Btu/hr, respectively. [Representative Rates at Which Heat and Moisture Are Given Off by People in Different States of Activity]

The sensible people load for the space is

$$q_{\text{people,sensible}} = A\left(\frac{\text{people}}{\text{area}}\right) q_{\text{sensible}}$$

$$= (40{,}000 \text{ ft}^2)\left(\frac{120 \text{ people}}{1000 \text{ ft}^2}\right)\left(250 \frac{\text{Btu}}{\text{hr} \cdot \text{person}}\right)$$

$$= 1{,}200{,}000 \text{ Btu/hr}$$

The latent people load for the space is

$$q_{people,latent} = A\left(\frac{people}{area}\right)q_{latent}$$

$$= (40{,}000 \text{ ft}^2)\left(\frac{120 \text{ people}}{1000 \text{ ft}^2}\right)\left(200 \frac{\text{Btu}}{\frac{\text{hr}}{\text{person}}}\right)$$

$$= 960{,}000 \text{ Btu/hr}$$

The sensible load for the space is

$$q_{sensible} = q_{lighting} + q_{equipment} + q_{people,sensible}$$

$$= 511{,}950 \frac{\text{Btu}}{\text{hr}} + 273{,}040 \frac{\text{Btu}}{\text{hr}}$$

$$+ 1{,}200{,}000 \frac{\text{Btu}}{\text{hr}}$$

$$= 1{,}984{,}990 \text{ Btu/hr}$$

The latent load for the space is

$$q_{latent} = q_{steam\,table} + q_{people,latent}$$

$$= 50{,}000 \frac{\text{Btu}}{\text{hr}} + 960{,}000 \frac{\text{Btu}}{\text{hr}}$$

$$= 1{,}010{,}000 \text{ Btu/hr}$$

The total load for the space is

$$q_{total} = q_{sensible} + q_{latent}$$

$$= 1{,}984{,}990 \frac{\text{Btu}}{\text{hr}} + 1{,}010{,}000 \frac{\text{Btu}}{\text{hr}}$$

$$= 2{,}994{,}990 \text{ Btu/hr} \quad (3{,}000{,}000 \text{ Btu/hr})$$

The answer is (D).

22. The throw of a diffuser is based on the distance traveled to the point where air velocity is reduced to 50 ft/min. The air distribution performance index (ADPI) method relies on the selection of a diffuser for which throw is related to the characteristic room dimension by published ratios, which are empirically derived for 9 ft ceiling heights.

For ceiling-mounted diffusers, the maximum ADPI is achieved for all cooling load densities when t_{50}/L is 0.8. [Air Diffusion Performance Index (ADPI) Selection Guide]

For the ceiling diffuser in this problem, the characteristic dimension, L, is the distance from the diffuser to the nearest wall, which is 6.0 ft. The throw of the diffuser is

$$\frac{t_{50}}{L} = 0.8$$
$$t_{50} = 0.8L$$
$$= (0.8)(6.0 \text{ ft})$$
$$= 4.8 \text{ ft} \quad (5.0 \text{ ft})$$

The answer is (A).

23. Using a psychrometric chart, find the humidity ratio for air at 70°F and 40% relative humidity. [ASHRAE Psychrometric Chart No. 1 - Normal Temperature at Sea Level]

$$\omega = 0.0062 \text{ lbm water/lbm air}$$

As the air in the car cools, the total mass of air and moisture in the car remains the same, making the process one of constant humidity. Using the psychrometric chart, start from the initial conditions, and proceed left on the chart along a constant humidity ratio line of 0.0062 lbm/lbm to 50°F. The final relative humidity is 80%.

The answer is (C).

24. According to the Darcy-Weisbach equation, the friction pressure loss in a conduit can be calculated as

Head Loss Due to Flow: Darcy-Weisbach Equation

$$h_f = f\frac{L}{D}\frac{v^2}{2g}$$

The hydraulic diameter, D_h, for noncircular piping is

Flow in Noncircular Conduits

$$D_h = \frac{4 \times \text{cross-sectional area of flowing fluid}}{\text{wetted perimeter}}$$

$$= \frac{(4)\left(\dfrac{18 \text{ in}}{12 \frac{\text{in}}{\text{ft}}}\right)\left(\dfrac{24 \text{ in}}{12 \frac{\text{in}}{\text{ft}}}\right)}{(2)\left(\dfrac{18 \text{ in}}{12 \frac{\text{in}}{\text{ft}}}\right) + (2)\left(\dfrac{24 \text{ in}}{12 \frac{\text{in}}{\text{ft}}}\right)}$$

$$= 1.714 \text{ ft}$$

The velocity in the duct is the flow divided by the cross-sectional area. [Measurement Relationships]

$$v = \frac{Q}{A} = \frac{\left(4500 \frac{ft^3}{min}\right)\left(12 \frac{in}{ft}\right)^2}{(18 \text{ in})(24 \text{ in})\left(60 \frac{sec}{min}\right)}$$

$$= 25 \text{ ft/sec}$$

The Darcy-Weisbach equation gives results in units of feet of the fluid of interest. However, it is traditional to measure fluid head of air in terms of inches of water, which requires additional unit conversions for length and density. To convert the head from inches of air to inches of water, multiply by the ratio of the densities in these two fluids, ρ_{air}/ρ_{water}. From a table of properties for air, the density of air at 68°F is 0.0752 lbm/ft^3. The density of water at standard conditions is 62.4 lbm/ft^3. [Properties of Water at Standard Conditions]

$$h_f = f\left(\frac{L}{D_h}\right)\left(\frac{v^2}{2g}\right)\left(\frac{\rho_{air}}{\rho_{water}}\right)$$

$$= (0.016)\left(\frac{240 \text{ ft}}{1.714 \text{ ft}}\right)\left(\frac{\left(25 \frac{ft}{sec}\right)^2}{(2)\left(32.2 \frac{ft}{sec^2}\right)}\right)$$

$$\times \left(12 \frac{in}{ft}\right)\left(\frac{0.0752 \frac{lbm \text{ air}}{ft^3}}{62.4 \frac{lbm \text{ water}}{ft^3}}\right)$$

$$= 0.31 \text{ in water} \quad (0.3 \text{ in water})$$

The answer is (B).

25. An energy balance must be established for the crawl space. The heat gain from the heated room above through the floor to the crawl space is equal to heat loss from the crawl space to the outdoor environment.

Heat Gain Through Interior Surfaces

$$q = UA(T_{adjacent \text{ space}} - T_{conditioned \text{ space}})$$

$$q_{floor} = q_{wall} + q_{inf}$$

The heat loss to the environment includes heat lost through the crawl space wall and heat lost warming up infiltrating air.

$$q_{floor} = q_{wall} + q_{inf}$$

$$U_{floor}A_{floor}\begin{pmatrix}T_{room}\\-T_{crawl \text{ space}}\end{pmatrix} = U_{wall}A_{wall}\begin{pmatrix}T_{crawl \text{ space}}\\-T_{outdoor}\end{pmatrix}$$

$$+ \dot{m}c_p\begin{pmatrix}T_{crawl \text{ space}}\\-T_{outdoor}\end{pmatrix}$$

To keep the pipes from freezing, the temperature of the crawl space, $T_{crawl \text{ space}}$, must remain above 32°F. Solve for the outdoor temperature, $T_{outdoor}$.

$$U_{floor}A_{floor}\begin{pmatrix}T_{room}\\-T_{crawl \text{ space}}\end{pmatrix} = U_{wall}A_{wall}T_{crawl \text{ space}}$$

$$- U_{wall}A_{wall}T_{outdoor}$$
$$+ \dot{m}c_p T_{crawl \text{ space}}$$
$$- \dot{m}c_p T_{outdoor}$$
$$= T_{crawl \text{ space}}(U_{wall}A_{wall} + \dot{m}c_p)$$
$$- T_{outdoor}(U_{wall}A_{wall} + \dot{m}c_p)$$

$$T_{outdoor}\begin{pmatrix}U_{wall}A_{wall}\\+\dot{m}c_p\end{pmatrix} = T_{crawl \text{ space}}(U_{wall}A_{wall} + \dot{m}c_p)$$

$$- U_{floor}A_{floor}$$
$$\times (T_{room} - T_{crawl \text{ space}})$$

$$T_{outdoor} = \frac{\begin{matrix}T_{crawl \text{ space}}(U_{wall}A_{wall} + \dot{m}c_p)\\- U_{floor}A_{floor}(T_{room} - T_{crawl \text{ space}})\end{matrix}}{U_{wall}A_{wall} + \dot{m}c_p}$$

$$= \frac{\begin{matrix}(32°F)\left(\left(0.15 \frac{Btu}{hr\text{-}ft^2\text{-}°F}\right)(320 \text{ ft}^2) + \left(900 \frac{ft^3}{hr}\right)\right.\\\left.\times \left(0.075 \frac{lbm}{ft^3}\right)\left(0.24 \frac{Btu}{lbm\text{-}°F}\right)\right)\\- \left(0.05 \frac{Btu}{hr\text{-}ft^2\text{-}°F}\right)(645 \text{ ft}^2)(72°F - 32°F)\end{matrix}}{\begin{matrix}\left(0.15 \frac{Btu}{hr\text{-}ft^2\text{-}°F}\right)(320 \text{ ft}^2) + \left(900 \frac{ft^3}{hr}\right)\\\times \left(0.075 \frac{lbm}{ft^3}\right)\left(0.24 \frac{Btu}{lbm\text{-}°F}\right)\end{matrix}}$$

$$= 11.9°F \quad (10°F)$$

The answer is (C).

26. From saturated water tables, at 180°F, h_{fg} is 989.86 Btu/lbm, and c_p is 1.00 Btu/lbm-°F. [Properties of Water at Standard Conditions] [Properties of Saturated Water and Steam (Temperature) - I-P Units]

The energy needed to evaporate the water is

$$E_{vapor} = m_{vapor} h_{fg}$$

An energy balance on the system gives

$$m_{vapor} h_{fg} = m_{remaining} c_p \Delta T_{remaining}$$

$$\Delta T_{remaining} = \frac{m_{vapor} h_{fg}}{m_{remaining} c_p}$$

8% of the water evaporates, so

$$\frac{m_{vapor}}{m_{remaining}} = \frac{8\%}{92\%} = 0.0870$$

The temperature change in the remaining water is

$$\Delta T_{remaining} = \left(\frac{m_{vapor}}{m_{remaining}}\right)\left(\frac{h_{fg}}{c_p}\right)$$

$$= (0.0870)\left(\frac{989.86 \frac{\text{Btu}}{\text{lbm}}}{1.00 \frac{\text{Btu}}{\text{lbm-°F}}}\right)$$

$$= 86.1°F$$

The remaining water has a temperature of

$$T_{remaining} = T_{initial} - \Delta T_{remaining}$$
$$= 180°F - 86.1°F$$
$$= 93.9°F \quad (94°F)$$

The answer is (C).

27. Find the brake horsepower required by the pump. [Commonly Used Equivalents] [Measurement Relationships]

Pump Power Equation

$$P_{pump} = \frac{Q\rho gh}{\eta_{pump}}$$

$$= \frac{Q\rho gh}{\eta_{pump} g_c}$$

$$= \frac{\left(6 \frac{\text{gal}}{\text{min}}\right)\left(8.34 \frac{\text{lbm}}{\text{gal}}\right)\left(32.2 \frac{\text{ft}}{\text{sec}^2}\right)(140 \text{ ft})}{(0.60)\left(32.2 \frac{\text{ft-lbm}}{\text{lbf-sec}^2}\right)\left(33{,}000 \frac{\text{ft-lbf}}{\text{hp-min}}\right)}$$

$$= 0.354 \text{ hp}$$

The solar array area is determined by dividing the power requirement by the product of the incident power available, $I_{incident}$, and the array collection efficiency, η_{array}.

$$A = \frac{P_{pump}}{\eta_{array} I_{incident}}$$

$$= \frac{(0.354 \text{ hp})\left(746 \frac{\text{W}}{\text{hp}}\right)}{(0.12)\left(28 \frac{\text{W}}{\text{ft}^2}\right)}$$

$$= 78.6 \text{ ft}^2 \quad (80 \text{ ft}^2)$$

The answer is (D).

28. The saturation efficiency for an air washer describes the extent to which the dry-bulb temperature is reduced to the theoretical minimum wet-bulb temperature.

Direct Evaporative Air Coolers

$$\varepsilon_e = (100)\frac{T_1 - T_2}{T_1 - T_s'}$$

T_1 is the dry-bulb entering, T_2 is the dry-bulb leaving, and T_s is the thermodynamic wet-bulb entering. The leaving dry-bulb temperature for this process is

$$T_2 = T_1 - \varepsilon_e(T_2 - T_s')$$
$$= 92°F - (0.84)(92°F - 57°F)$$
$$= 62.6°F$$

The sensible precooling benefit is

$$q_{sensible} = \dot{m} c_p (T_1 - T_2)$$
$$= \dot{V} \rho c_p (T_1 - T_2)$$
$$= \left(20{,}000 \frac{\text{ft}^3}{\text{min}}\right)\left(60 \frac{\text{min}}{\text{hr}}\right)\left(0.075 \frac{\text{lbm}}{\text{ft}^3}\right)$$
$$\times \left(0.24 \frac{\text{Btu}}{\text{lbm-°F}}\right)(92°F - 62.6°F)$$
$$= 635{,}040 \text{ Btu/hr}$$

Chiller loads are typically expressed in units of tons of refrigeration. [Measurement Relationships]

$$q_{sensible} = \frac{635{,}040 \frac{\text{Btu}}{\text{hr}}}{12{,}000 \frac{\text{Btu}}{\text{hr-ton}}} = 52.92 \text{ tons} \quad (53 \text{ tons})$$

The answer is (C).

29. The only load is the heat loss through the brick wall. The heat loss can be calculated as

Composite Plane Wall
$$q_{\text{wall}} = U_{\text{wall}} A_{\text{wall}} (T_{\text{inside}} - T_{\text{outside}})$$

The overall heat transfer coefficient (U-factor) is calculated by taking the reciprocal of the total thermal resistance of the wall assembly, including air surfaces. If the construction of the wall is thermally homogenous (continuous without conductive penetrations), resistance of the wall is the sum of the individual resistances. Individual resistances can be obtained from a wide variety of sources, the most comprehensive being the *ASHRAE Handbook—Fundamentals*. Typical resistance values for the individual wall components are listed in the following table. [Thermal Resistance of Building Materials] [Surface Film Coefficients/Resistances for Air]

wall components	resistance $\left(\dfrac{\text{hr-ft}^2\text{-}°\text{F}}{\text{Btu}}\right)$
8 in brick	0.80 (0.10 per inch thickness)
2 in polystyrene	8.00 (4 per inch thickness)
5/8 in gypsum board	0.56
outdoor air film	0.17 (winter value)
indoor air film (vertical)	0.68 (heat flow direction horizontal)
total resistance	10.21

The wall's overall heat transfer coefficient, U_{wall}, is

Composite Plane Wall
$$U_{\text{wall}} = \frac{1}{R_{\text{wall}}}$$
$$= \frac{1}{10.21\ \dfrac{\text{hr-ft}^2\text{-}°\text{F}}{\text{Btu}}}$$
$$= 0.098\ \text{Btu/hr-ft}^2\text{-}°\text{F}$$

The heat loss through the wall is

Composite Plane Wall
$$q_{\text{wall}} = U_{\text{wall}} A_{\text{wall}} (T_{\text{inside}} - T_{\text{outside}})$$
$$= \left(0.098\ \frac{\text{Btu}}{\text{hr-ft}^2\text{-}°\text{F}}\right)(800\ \text{ft}^2)(74°\text{F} - 10°\text{F})$$
$$= 5018\ \text{Btu/hr}$$

Find the heating capacity. [Measurement Relationships]

$$q_{\text{heater}} = q_{\text{wall}}(1 + \text{SF})$$
$$q_{\text{heater}} = \frac{\left(5018\ \dfrac{\text{Btu}}{\text{hr}}\right)(1.0 + 0.25)}{3412\ \dfrac{\text{Btu}}{\text{kW-hr}}}$$
$$= 1.84\ \text{kW}\quad (1.9\ \text{kW})$$

The answer is (B).

30. The sensible capacity of the coil can be calculated as

$$q_{\text{sensible}} = \dot{m} c_p (T_{\text{entering,db}} - T_{\text{leaving,db}})$$
$$= Q \rho c_p (T_{\text{db,air,entering}} - T_{\text{db,air,leaving}})$$

The airflow through the coil is the product of the velocity of the air and the cross-sectional area of the coil.

Continuity Equation
$$Q = A\text{v} = (3\ \text{ft})(4\ \text{ft})\left(450\ \frac{\text{ft}}{\text{min}}\right)$$
$$= 5400\ \text{ft}^3/\text{min}$$

The sensible capacity of the coil is

$$q_{\text{sensible}} = Q \rho c_p (T_{\text{db,air,entering}} - T_{\text{db,air,leaving}})$$
$$= \left(5400\ \frac{\text{ft}^3}{\text{min}}\right)\left(60\ \frac{\text{min}}{\text{hr}}\right)\left(0.075\ \frac{\text{lbm}}{\text{ft}^3}\right)$$
$$\quad\times \left(0.24\ \frac{\text{Btu}}{\text{lbm-}°\text{F}}\right)(80°\text{F} - 56°\text{F})$$
$$= 139{,}968\ \text{Btu/hr}$$

The sensible heat ratio of the coil is the ratio of the sensible capacity to the total capacity.

Psychrometric Properties
$$\text{SHR} = \frac{\text{sensible heat gain}}{\text{total heat gain}} = \frac{q_{\text{sensible}}}{q_{\text{total}}}$$

The total coil capacity is

$$q_{\text{total}} = \frac{q_{\text{sensible}}}{\text{SHR}} = \frac{139{,}968\ \dfrac{\text{Btu}}{\text{hr}}}{0.70}$$
$$= 199{,}954\ \text{Btu/hr}\quad (200{,}000\ \text{Btu/hr})$$

The answer is (D).

31. The coefficient of performance (COP) is the ratio of useful cooling to the power the compressor requires to accomplish the cooling.

Efficiency

$$\text{COP} = \frac{\text{capacity}}{\text{input}}$$
$$= \frac{q_{\text{cooling}}}{P}$$

The load rejected to the cooling tower includes both the heat rejected as cooling load and the required compressor energy. Find the total load rejected to the cooling tower. [Measurement Relationships]

$$q_{\text{condenser}} = \dot{m} c_p \Delta T_{\text{condenser}}$$

$$= \frac{\left(2530\, \frac{\text{gal}}{\text{min}}\right)\left(8.34\, \frac{\text{lbm}}{\text{gal}}\right)\left(60\, \frac{\text{min}}{\text{hr}}\right) \times \left(1.0\, \frac{\text{Btu}}{\text{lbm-}°\text{F}}\right)(92°\text{F} - 83°\text{F})}{12{,}000\, \frac{\text{Btu}}{\text{hr-ton}}}$$

$$= 949.5\ \text{tons}\quad (950\ \text{tons})$$

The compressor heat is the difference between the total load rejected to the condenser and the cooling capacity of the chiller.

$$q_{\text{compressor}} = q_{\text{rejected}} - q_{\text{chiller}}$$
$$= 950\ \text{tons} - 840\ \text{tons}$$
$$= 110\ \text{tons}$$

The chiller COP is

$$\text{COP} = \frac{q_{\text{chiller}}}{q_{\text{compressor}}} = \frac{840\ \text{tons}}{110\ \text{tons}}$$
$$= 7.67\quad (8)$$

The answer is (D).

32. Convert the temperature from degrees Fahrenheit to Rankine. Obtain the gas constant, R, from a table. [Thermal and Physical Properties of Ideal Gases (at Room Temperature)]

$$T_{°\text{R}} = T_{°\text{F}} + 460° = 100°\text{F} + 460°$$
$$= 560°\text{R}$$
$$R_{\text{nitrogen}} = 55.16\ \text{ft-lbf/lbm-}°\text{R}$$

Since the pressure change is only due to the release of mass, it is reasonable to assume constant temperature. The initial mass, m_1, is 5 lbm. Rearrange the ideal gas law and calculate the mass remaining in the bottle, m_2.

Ideal Gas

$$pV = m_2 RT$$
$$m_2 = \frac{pV}{RT}$$
$$= \frac{\left(150\, \frac{\text{lbf}}{\text{in}^2}\right)\left(12\, \frac{\text{in}}{\text{ft}}\right)^2 (6\ \text{ft}^3)}{\left(55.16\, \frac{\text{ft-lbf}}{\text{lbm-}°\text{R}}\right)(560°\text{R})}$$
$$= 4.19\ \text{lbm}$$

Find the mass of gas released.

$$m_{\text{released}} = m_1 - m_2 = 5\ \text{lbm} - 4.19\ \text{lbm}$$
$$= 0.81\ \text{lbm}\quad (0.80\ \text{lbm})$$

The answer is (B).

33. Airfoil fans have the highest efficiency due to the airfoil contour of each of their blades. The fans with the next highest efficiency are backward curved and backward inclined. Forward-curved fans have the lowest efficiency.

The answer is (D).

34. The coil load due to the ventilation air is the rate at which sensible and latent heat must be removed from the ventilation air to reduce it to the return state.

Moist-Air Sensible Heating or Cooling

$$q_{\text{total}} = \dot{m}\Delta h = Q\rho(h_{\text{outdoor}} - h_{\text{supply}})$$

From the psychrometric chart, at 94°F dry-bulb and 72°F wet-bulb, the enthalpy of the outside air, h_{outdoor}, is 35.6 Btu/lbm. [ASHRAE Psychrometric Chart No. 1 - Normal Temperature at Sea Level]

At 55°F and 30% relative humidity,

$$h_{\text{supply}} = 16.2\ \text{Btu/lbm}$$

Interpolating from a table of properties of air at atmospheric pressure, for air at 55°F, the density is 0.0769 lbm/ft³. [Properties of Air at Atmospheric Pressure]

The coil load is

$$q_{total} = Q\rho(h_{outdoor} - h_{supply})$$
$$= \left(850 \ \frac{ft^3}{min}\right)\left(60 \ \frac{min}{hr}\right)\left(0.0769 \ \frac{lbm}{ft^3}\right)$$
$$\times \left(35.6 \ \frac{Btu}{lbm} - 16.2 \ \frac{Btu}{lbm}\right)$$
$$= 75,495 \ Btu/hr \quad (80,000 \ Btu/hr)$$

The answer is (C).

35. The reheat coil must have sufficient heating capacity to offset both the winter space heat loss and the colder supply air.

$$q_{coil} = q_{space} + q_{reheat}$$

The reheat load is given by

$$q_{reheat} = \dot{m}c_p(T_{room} - T_{supply})$$
$$= Q_{min}\rho_{air}c_p(T_{room} - T_{supply})$$

The lowest airflow is the product of the minimum stop fraction and the maximum flow.

$$Q_{min} = F_{min}Q_{max} = (0.30)\left(2400 \ \frac{ft^3}{min}\right)$$
$$= 720 \ ft^3/min$$

The reset chart indicates that the supply air temperature will be reset to 64°F when the outdoor temperature is 10°F.

$$q_{reheat} = Q_{min}\rho_{air}c_p(T_{room} - T_{supply})$$
$$= \left(720 \ \frac{ft^3}{min}\right)\left(60 \ \frac{min}{hr}\right)\left(0.075 \ \frac{lbm}{ft^3}\right)$$
$$\times \left(0.24 \ \frac{Btu}{lbm\text{-}°F}\right)(72°F - 64°F)$$
$$= 6220 \ Btu/hr$$

The total heating coil load is

$$q_{coil} = q_{space} + q_{reheat}$$
$$= 45,000 \ \frac{Btu}{hr} + 6220 \ \frac{Btu}{hr}$$
$$= 51,220 \ Btu/hr \quad (51,000 \ Btu/hr)$$

The answer is (C).

36. The time required for the steam to cool to the ambient temperature is determined by dividing the energy reduction (from 212°F saturated steam to 60°F water) by the rate of heat loss from the pipe.

$$t = \frac{E_{initial} - E_{final}}{q_{pipe}}$$

From standard pipe dimension schedules, the interior diameter of a 2 in schedule-40 pipe is 2.067 in. [Schedule 40 Steel Pipe]

The volume of the steam in the pipe is

$$V = A_i L$$
$$= \frac{\pi}{4}D^2 L$$
$$= \left(\frac{\pi}{4}\right)\left(\frac{2.067 \ in}{12 \ \frac{in}{ft}}\right)^2 (80 \ ft)$$
$$= 1.864 \ ft^3$$

From saturated steam tables, the specific volume of saturated 212°F steam is 26.78 ft³/lbm. [Properties of Saturated Water and Steam (Temperature) - I-P Units]

The mass of the steam in the pipe is

$$m_{steam} = \rho V = \frac{V}{v}$$
$$= \frac{1.864 \ ft^3}{26.78 \ \frac{ft^3}{lbm}}$$
$$= 0.0696 \ lbm$$

The initial enthalpy of the saturated 212°F steam is 1150.30 Btu/lbm. [Properties of Saturated Water and Steam (Temperature) - I-P Units]

Since the steam is initially saturated, any cooling will result in some condensation. At 60°F, all the steam will have condensed, and there will be 0.0696 lbm of 60°F liquid water in the pipe. The enthalpy of 60°F water is approximately 28.08 Btu/lbm. [Properties of Saturated Water and Steam (Temperature) - I-P Units]

The total energy loss is

$$\Delta E_{steam} = E_{initial} - E_{final}$$
$$= m_{steam}\Delta h$$
$$= (0.0696 \ lbm)\left(1150.3 \ \frac{Btu}{lbm} - 28.08 \ \frac{Btu}{lbm}\right)$$
$$= 78.11 \ Btu$$

The time it takes the steam to cool is

$$t = \frac{\Delta E_{\text{steam}}}{q_{\text{pipe}}}$$

$$= \frac{(78.11 \text{ Btu})\left(60 \, \frac{\text{min}}{\text{hr}}\right)}{6200 \, \frac{\text{Btu}}{\text{hr}}}$$

$$= 0.756 \text{ min} \quad (1 \text{ min})$$

The answer is (B).

37. From steam tables, the entropy of the steam entering the turbine at 1200°F and 700 psia, s_3, is 1.769 Btu/lbm-°R. [Properties of Superheated Steam - I-P Units]

For maximum efficiency, the entropy of the steam entering the condenser, s^4, has the same value. The quality can be found from the equation

Properties for Two-Phase (Vapor-Liquid) Systems

$$s_3 = s_4 = s_f + xs_{fg}$$

$$x = \frac{s_3 - s_f}{s_g - s_f}$$

s_f and s_g are the entropies of the saturated liquid and vapor, respectively, and are taken from a saturated steam table for a pressure of 2 psia. [Properties of Saturated Water and Steam (Pressure) - I-P Units]

$$x = \frac{s_3 - s_f}{s_g - s_f} = \frac{1.769 \, \frac{\text{Btu}}{\text{lbm-°R}} - 0.1749 \, \frac{\text{Btu}}{\text{lbm-°R}}}{1.9195 \, \frac{\text{Btu}}{\text{lbm-°R}} - 0.1749 \, \frac{\text{Btu}}{\text{lbm-°R}}}$$

$$= 0.9137$$

The enthalpy of the steam entering the condenser is

Properties for Two-Phase (Vapor-Liquid) Systems

$$h_4 = h_f + xh_{fg}$$

h_f and h_g are the enthalpies of the saturated liquid and vapor, respectively, and are taken from a saturated steam table for a pressure of 2 psia. [Properties of Saturated Water and Steam (Pressure) - I-P Units]

$$h_4 = h_f + x(h_g - h_f)$$

$$= 94.00 \, \frac{\text{Btu}}{\text{lbm}} + (0.9137)\left(1115.74 \, \frac{\text{Btu}}{\text{lbm}} - 94.00 \, \frac{\text{Btu}}{\text{lbm}}\right)$$

$$= 1027.6 \text{ Btu/lbm}$$

From steam tables, the enthalpy between the turbine and boiler, h_3, for 1200°F and 700 psia is 1625.9 Btu/lbm. The enthalpy of the saturated 2 psia liquid, h_1, is 94.00 Btu/lbm. [Properties of Superheated Steam - I-P Units]

The efficiency of the cycle is

$$\eta = \frac{W}{Q}$$

$$= \frac{h_3 - h_4}{h_3 - h_1}$$

$$= \frac{1625.9 \, \frac{\text{Btu}}{\text{lbm}} - 1027.6 \, \frac{\text{Btu}}{\text{lbm}}}{1625.9 \, \frac{\text{Btu}}{\text{lbm}} - 94.00 \, \frac{\text{Btu}}{\text{lbm}}}$$

$$= 0.391 \quad (39\%)$$

The answer is (C).

38. Since the temperatures of water entering and leaving the condenser are fixed, the required flow is determined by the total load to be rejected, which includes both the load absorbed by the evaporator and the heat added by the compressor. The relationship for the condenser water temperature rise is

$$q_{\text{rejected}} = \dot{m}_{\text{condenser, water}} c_p \Delta T$$

The ΔT term represents the water temperature change, and the load to be rejected is

$$q_{\text{rejected}} = q_{\text{refrigeration}} + q_{\text{comp, heat}}$$

The required condenser water can be determined as

$$\dot{m}_{\text{condenser, water}} = \frac{q_{\text{refrigeration}} + q_{\text{comp, heat}}}{c_p \Delta T}$$

The refrigeration capacity of the chiller is the product of the refrigerant mass flow rate and the refrigeration effect.

$$q_{\text{refrigeration}} = \dot{m}_{\text{refrigerant}}(h_{\text{evap, leaving}} - h_{\text{evap, entering}})$$
$$= \dot{m}_{\text{refrigerant}}(h_{\text{comp, entering}} - h_{\text{evap, entering}})$$

The enthalpy of the refrigerant mixture entering the chiller evaporator is equal to that of the saturated liquid leaving the chiller condenser since there is no change in enthalpy when the refrigerant goes through the throttling valve. Refrigerant enthalpies are found in an R-22 property table. Find the enthalpy of the saturated liquid at 90°F. [Pressure Versus Enthalpy Curves for Refrigerant 22]

$$h_{\text{evap, entering}} = 36.12 \text{ Btu/lbm}$$

Since there is no superheat, the enthalpy of the refrigerant entering the compressor is equal to the enthalpy of the saturated vapor at 10°F. The enthalpy of the saturated vapor at 10°F is

$$h_{\text{comp, entering}} = 105.27 \text{ Btu/lbm}$$

The useful refrigeration is

$$\begin{aligned}q_{\text{refrigeration}} &= \dot{m}_{\text{refrigerant}}(h_{\text{comp, entering}} - h_{\text{evap, entering}}) \\ &= \left(117{,}000 \, \frac{\text{lbm}}{\text{hr}}\right)\left(105.27 \, \frac{\text{Btu}}{\text{lbm}} - 36.12 \, \frac{\text{Btu}}{\text{lbm}}\right) \\ &= 8{,}090{,}433 \text{ Btu/hr}\end{aligned}$$

Based on the definition of coefficient of performance (COP), the compressor work can be determined.

$$\begin{aligned}q_{\text{comp}} &= \frac{q_{\text{refrigeration}}}{\text{COP}} \\ &= \frac{8{,}090{,}433 \, \frac{\text{Btu}}{\text{hr}}}{5.5} \\ &= 1{,}470{,}988 \text{ Btu/hr}\end{aligned}$$

The required rate of heat removal is

$$\begin{aligned}q_{\text{rejected}} &= q_{\text{refrigeration}} + q_{\text{comp}} \\ &= 8{,}090{,}433 \, \frac{\text{Btu}}{\text{hr}} + 1{,}470{,}988 \, \frac{\text{Btu}}{\text{hr}} \\ &= 9{,}561{,}421 \text{ Btu/hr}\end{aligned}$$

Find the condenser water flow that is needed to remove the heat. [Measurement Relationships]

$$\begin{aligned}\dot{m}_{\text{condenser}} &= \rho Q_{\text{condenser}} \\ &= \frac{q_{\text{rejected}}}{c_p \Delta T} \\ Q_{\text{condenser}} &= \frac{q_{\text{rejected}}}{\rho c_p \Delta T} \\ &= \frac{9{,}561{,}421 \, \frac{\text{Btu}}{\text{hr}}}{\left(8.34 \, \frac{\text{lbm}}{\text{gal}}\right)\left(1.0 \, \frac{\text{Btu}}{\text{lbm-°F}}\right)} \\ &\quad \times (95°\text{F} - 85°\text{F})\left(60 \, \frac{\text{min}}{\text{hr}}\right) \\ &= 1910.8 \text{ gal/min} \quad (1900 \text{ gpm})\end{aligned}$$

The answer is (C).

39. Use the equation for the volumetric flow rate

Continuity Equation

$$Q = Av$$

Rewrite the equation using the area of a circle, and solve for the pipe radius. [Measurement Relationships]

$$\begin{aligned}Q &= \pi r^2 v \\ r &= \sqrt{\frac{Q}{\pi v}} \\ &= \sqrt{\frac{\left(2200 \, \frac{\text{gal}}{\text{min}}\right)\left(12 \, \frac{\text{in}}{\text{ft}}\right)^3}{\pi \left(7.48 \, \frac{\text{gal}}{\text{ft}^3}\right)\left(60 \, \frac{\text{sec}}{\text{min}}\right)\left(5 \, \frac{\text{ft}}{\text{sec}}\right)\left(12 \, \frac{\text{in}}{\text{ft}}\right)}} \\ &= 6.7 \text{ in}\end{aligned}$$

The diameter is

$$\begin{aligned}D &= 2r = (2)(6.7 \text{ in}) \\ &= 13.4 \text{ in}\end{aligned}$$

The minimum pipe diameter needed to obtain the flow rate desired is 14 in.

The answer is (C).

40. The temperature of the air leaving the coil is determined by subtracting the cooling coil's effect on temperature from the dry-bulb temperature of the air entering the coil.

$$T_{\text{coil, leaving}} = T_{\text{coil, entering}} - \Delta T_{\text{coil}}$$

Find the temperature of the air entering the cooling coil which is simply the mixed air dry-bulb temperature.

$$\begin{aligned}T_{\text{coil, entering}} &= T_{\text{return}} + \left(\frac{Q_{\text{outside}}}{Q_{\text{return}} + Q_{\text{outside}}}\right) \\ &\quad \times (T_{\text{outside}} - T_{\text{return}}) \\ &= 75°\text{F} + \left(\frac{2300 \, \frac{\text{ft}^3}{\text{min}}}{7000 \, \frac{\text{ft}^3}{\text{min}} + 2300 \, \frac{\text{ft}^3}{\text{min}}}\right) \\ &\quad \times (90°\text{F} - 75°\text{F}) \\ &= 78.7°\text{F}\end{aligned}$$

The change in temperature across the cooling coil is determined by the sensible heat removed from the air stream. The specific heat of air is 0.24 Btu/lbm-°F. [Heat Gain Calculations Using Standard Air Values]

The sensible heating relationship is $q_{sensible} = \dot{m} c_p \Delta T$.

$$\Delta T_{coil} = \frac{q_{sensible}}{\dot{m} c_p} = \frac{q_{sensible}}{Q \rho c_p}$$

$$= \frac{235{,}000 \, \frac{Btu}{hr}}{\left(7000 \, \frac{ft^3}{min} + 2300 \, \frac{ft^3}{min}\right)\left(0.075 \, \frac{lbm}{ft^3}\right)}$$
$$\times \left(0.24 \, \frac{Btu}{lbm\text{-}°F}\right)\left(60 \, \frac{min}{hr}\right)$$
$$= 23.4°F$$

The air temperature leaving the coil is

$$T_{coil, leaving} = T_{coil, entering} - \Delta T_{coil} = 78.7°F - 23.4°F$$
$$= 55.3°F \quad (55°F)$$

The answer is (B).

41. Referring to the diagram, process 1-2 moves to the left in the direction of sensible cooling and down in the direction of sensible cooling. Process 3-4 is just the opposite, although with more sensible heating. Process 4-6 uses evaporating sweat to remove heat, which is the same process the human body uses to cool down. Process 4-5 involves no change in temperature; it is just removing the moisture out of the air. [ASHRAE Psychrometric Chart No. 1 - Normal Temperature at Sea Level] [Moist-Air Sensible Heating or Cooling] [Moist-Air Cooling and Dehumidification]

The answer is (D).

42. Note the nonstandard units used by the code. The four air quantities are calculated to determine which one will govern. The negative pressure airflow requirement is

$$Q_1 = 2610 A_e \sqrt{\Delta p} = (2610)(1.5 \, ft^2)\sqrt{0.05 \, in \, water}$$
$$= 875 \, ft^3/min$$

The gross floor area airflow requirement is

$$Q_2 = 0.5 A_{gf} = (0.5)(1600 \, ft^2) = 800 \, ft^3/min$$

The requirement for the temperature rise limitation airflow is

$$Q_3 = \frac{\sum q}{1.08 \Delta T} = \frac{15{,}000 \, \frac{Btu}{hr}}{(1.08)(104°F - 90°F)} = 992 \, ft^3/min$$

The requirement for the emergency purge airflow is

$$Q_4 = 100\sqrt{m} = 100\sqrt{150 \, lbm}$$
$$= 1225 \, ft^3/min \quad (1200 \, ft^3/min)$$

The emergency purge requirement sets the requirement for the room's dedicated mechanical exhaust.

The answer is (D).

43. The first law for an open system at steady state is

Steady-Flow Systems

$$\sum \dot{m}_i \left(h_i + \frac{v_i^2}{2} + gZ_i \right)$$
$$- \sum \dot{m}_e \left(h_e + \frac{v_e^2}{2} + gZ_e \right) + \dot{Q}_{in} - \dot{W}_{out} = 0$$

The condenser has two inlets (water and steam) and two outlets (water and condensate). The condenser does no work. Although heat transfers from the steam stream to the water stream within the condenser, no heat is assumed to transfer between the condenser and the environment. The contributions from velocity and elevation are negligible.

$$\sum \dot{m}_i (h_i) - \sum \dot{m}_e (h_e) = 0$$
$$\dot{m}_{steam}(h_{steam,i}) + \dot{m}_{water}(h_{water,i})$$
$$- \dot{m}_{steam}(h_{steam,e}) - \dot{m}(h_{water,e}) = 0$$
$$\dot{m}_{steam}(h_{steam,i} - h_{steam,e}) = \dot{m}_{water}(h_{water,e} - h_{water,i})$$

The water does not change phase and is at constant pressure.

$$\dot{m}_{water}(h_{water,e} - h_{water,i}) = \dot{m}_{water} c_{p,water}(T_{water,e} - T_{water,i})$$

Substituting into the previous equation and solving for the mass flow rate of the water gives

$$\dot{m}_{steam}(h_{steam,i} - h_{steam,e}) = \dot{m}_{water} c_{p,water}(T_{water,e} - T_{water,i})$$
$$\dot{m}_{water} = \frac{\dot{m}_{steam}(h_{steam,i} - h_{steam,e})}{c_{p,water}(T_{water,e} - T_{water,i})}$$

From saturated water tables, for a pressure of 4 psia, the enthalpy of the vapor entering the condenser, $h_{steam,i}$, and the enthalpy of the saturated liquid exiting the condenser, $h_{steam,e}$, is 1126.91 Btu/lbm and 120.87 Btu/lbm, respectively. [**Properties of Saturated Water and Steam (Pressure) - I-P Units**]

For water at 65°F, the specific heat of water is approximately 1.0 Btu/lbm-°F [**Properties of Water at Standard Conditions**]

The mass flow rate of the exiting water is

$$\dot{m}_{\text{water}} = \frac{\dot{m}_{\text{steam}}(h_{\text{steam},i} - h_{\text{steam},e})}{c_{p,\text{water}}(T_{\text{water},e} - T_{\text{water},i})}$$

$$= \frac{\left(2200 \frac{\text{lbm}}{\text{hr}}\right)\left(1126.91 \frac{\text{Btu}}{\text{lbm}} - 120.87 \frac{\text{Btu}}{\text{lbm}}\right)}{\left(1.0 \frac{\text{Btu}}{\text{lbm -°F}}\right)(95°\text{F} - 65°\text{F})}$$

$$= 73{,}776 \text{ lbm/hr} \quad (74{,}000 \text{ lbm/hr})$$

The answer is (C).

44. The heat transfer to the soil is determined by the following equation.

Fourier's Law of Conduction

$$q = -kA\frac{dT}{dx}$$

The heat transfer from the pipes is determined by the following equation.

Water-Cooled Condensers

$$Q = \frac{q_o}{\rho c_p(T_2 - T_1)}$$

$$q_o = \rho Q c_p(T_2 - T_1)$$

Since q_o and q are equivalent,

$$kA\frac{dT}{dx} = \rho Q c_p(T_2 - T_1)$$

The heat transfer to the soil must remain constant.

$$kA\frac{dT}{dx} = \rho Q c_p(T_2 - T_1) = \text{constant}$$

The increase in moisture content in the soil will decrease the thermal conductivity of the soil. In order to maintain constant heat transfer, the flow rate must also decrease. Therefore, the flow rate of the water must be decreased.

The answer is (B).

45. The thermal energy that can be stored in the slab depends on the mass, specific heat, and the temperature difference between the slab and the room. The temperature variation of the slab with time is

Constant Fluid Temperature

$$T - T_\infty = (T_i - T_\infty)e^{-\beta t}$$

Calculate the time constant, β.

Constant Fluid Temperature

$$\beta = \frac{hA_s}{\rho V c_p} = \frac{\left(16.8 \frac{\text{Btu}}{\text{hr-ft}^2\text{-°F}}\right)(20 \text{ ft})(30 \text{ ft})}{\left(140 \frac{\text{lbm}}{\text{ft}^3}\right)(0.50 \text{ ft})(20 \text{ ft})}$$

$$\times (30 \text{ ft})\left(0.22 \frac{\text{Btu}}{\text{lbm-°F}}\right)$$

$$= 1.091 \text{ hr}^{-1}$$

Rearrange the equation for temperature variation so that the body temperature, T, represents the temperature as a function of time.

$$T(t) = (T_i - T_\infty)e^{-\beta t} + T_\infty$$

The temperature after 1 hr is

$$T(1 \text{ hr}) = (T_i - T_\infty)e^{-\beta t} + T_\infty$$

$$= (82°\text{F} - 60°\text{F})e^{(-1.091 \text{ hr}^{-1})(1 \text{ hr})} + 60°\text{F}$$

$$= 67.39°\text{F} \quad (67.4°\text{F})$$

To find the temperature change in the slab, take the difference between the initial temperature and the temperature after 1 hr.

$$T_i - T(1 \text{ hr}) = 82°\text{F} - 67.4°\text{F} = 14.6°\text{F} \quad (15°\text{F})$$

The answer is (D).

46. According to the ASHRAE guidelines cited, the recommended storage tank provides an hour of probable demand, appropriately factored for the facility type.

$$V_{\text{tank}} = V_{\text{probable}} F_{\text{storage factor}}$$

The probable hot-water demand is determined by adjusting the maximum demand by an industry demand factor appropriate to the facility type.

$$Q_{\text{probable}} = Q_{\text{maximum possible}} F_{\text{demand}}$$

The maximum possible demand is most easily determined in a tabular calculation format.

type of fixture	total no. of fixtures	demand per fixture (gal/hr)	total demand per fixture type (gal/hr)
lavatory	44	2	88
bathtub	22	20	440
shower	22	30	660
kitchen sink	22	10	220
dishwasher	22	15	330
maximum possible demand			1738

The maximum probable demand is

$$Q_{\text{probable}} = Q_{\text{maximum possible}} F_{\text{demand}}$$
$$= \left(1738 \frac{\text{gal}}{\text{hr}}\right)(0.30)$$
$$= 521.4 \text{ gal/hr}$$

The recommended volume of the storage tank is

$$V_{\text{tank}} = V_{\text{probable}} F_{\text{storage factor}}$$
$$= Q_{\text{probable}} t F_{\text{storage factor}}$$
$$= \left(521.4 \frac{\text{gal}}{\text{hr}}\right)(1 \text{ hr})(1.25)$$
$$= 651.8 \text{ gal} \quad (650 \text{ gal})$$

The answer is (B).

47. For isentropic processes, the entropy remains constant ($\Delta s = 0$).

The specific work done on an isentropic process is

Special Cases of Closed Systems (With No Change in Kinetic or Potential Energy)

$$w = \frac{R(T_2 - T_1)}{1 - k}$$

From a table of thermal and physical properties of ideal gases (at room temperature), the gas constant for air is 53.35 ft-lbf/lbm-°R. [Thermal and Physical Properties of Ideal Gases (at Room Temperature)]

Converting the temperature of the air, T_1, to Rankine gives

$$T_1 = 65°\text{F} + 460° = 525°\text{R}$$

Find the ratio of specific heats, k.

Mach Number

$$k = \frac{c_p}{c_v}$$
$$= \frac{0.240 \dfrac{\text{Btu}}{\text{lbm-}°\text{R}}}{0.171 \dfrac{\text{Btu}}{\text{lbm-}°\text{R}}}$$
$$= 1.4$$

An isentropic compression process has the following relationship.

Ideal Gas

$$\frac{T_2}{T_1} = \left(\frac{p_2}{p_1}\right)^{\frac{k-1}{k}}$$

$$T_2 = T_1\left(\frac{p_2}{p_1}\right)^{\frac{k-1}{k}}$$
$$= (525°\text{R})\left(\frac{720 \text{ psia}}{14.7 \text{ psia}}\right)^{\frac{1.4-0.4}{1.4}}$$
$$= 1596°\text{R}$$

Determine the specific work.

$$w = \frac{R(T_2 - T_1)}{1 - k} = \frac{\left(53.3 \dfrac{\text{ft-lbf}}{\text{lbm-}°\text{R}}\right)(1596°\text{R} - 525°\text{R})}{1 - 1.4}$$
$$= 142{,}705 \text{ ft-lbf/lbm}$$

Multiply the specific work by the mass of air to determine the needed work.

$$W = mw$$
$$= (10 \text{ lbm})\left(142{,}705 \frac{\text{ft-lbf}}{\text{lbm}}\right)\left(\frac{1 \text{ Btu}}{778 \dfrac{\text{ft-lbf}}{\text{lbm}}}\right)$$
$$= 1830 \text{ Btu} \quad (1800 \text{ Btu})$$

The answer is (D).

48. From the table, the incident total irradiance for an east-facing window in May at 0800 is

$$E_t = 218 \text{ Btu/hr-ft}^2$$

The solar heat gain coefficient, SHGC, is given as 0.87, and the overall heat transfer coefficient, U, is 1.2 Btu/hr-ft²-°F. The instantaneous heat gain is

Fenestration

$$q = UA_{\text{pf}}(T_{\text{out}} - T_{\text{in}}) + (\text{SHGC})A_{\text{pf}}E_T$$
$$\quad + C(\text{AL})A_{\text{pf}}\rho C_p(T_{\text{out}} - T_{\text{in}})$$

No air leakage is given in the problem; assume it to be 0. The total instantaneous heat gain for the window is

$$q = UA_{pf}(T_{out} - T_{in}) + (SHGC)A_{pf}E_T$$

$$= \left(1.2 \frac{Btu}{hr\text{-}ft^2\text{-}°F}\right)(12\ ft^2)(42°F - 75°F)$$

$$+ (0.87)(12\ ft^2)\left(218 \frac{Btu}{hr\text{-}ft^2}\right)$$

$$= 1800.7\ Btu/hr \quad (1800\ Btu/hr)$$

The answer is (B).

49. The heat loss from conduction through the wall is

Composite Plane Wall

$$q = UA(T_{inside} - T_{outside})$$

The difference in heat loss from conduction between the original temperature of 70°F and the new temperature of 58°F is

$$q_{70°F} - q_{58°F} = UA(70°F - T_{outside}) - UA(58°F - T_{outside})$$
$$= UA(70°F - 58°F)$$
$$= UA(12°F)$$

The total heat loss from walls, windows, and the roof is

$$q_{70°F} - q_{58°F} = \left(0.15 \frac{Btu}{hr\text{-}ft^2\text{-}°F}\right)(10{,}000\ ft^2)(12°F)$$

$$+ \left(1.10 \frac{Btu}{hr\text{-}ft^2\text{-}°F}\right)(2500\ ft^2)(12°F)$$

$$+ \left(0.06 \frac{Btu}{hr\ ft^2\ °F}\right)(25{,}000)(12°F)$$

$$= 69{,}000\ Btu/hr$$

Find the heat loss from air infiltration during ventilation air exchange.

$$q_{infiltration} = 0.018 \frac{Btu}{°F} QV\Delta T$$

$$\begin{aligned} q_{infiltration,70°F} \\ - q_{infiltration,58°F} \end{aligned} = 0.018\ QV(70°F - T_{outside})$$

$$-0.018 QV(58°F - T_{outside})$$
$$= 0.018QV(12°F)$$
$$= \left(0.018 \frac{Btu}{°F}\right)\left(0.5 \frac{air\ change}{hr}\right)$$
$$\times \left(600{,}000 \frac{ft^3}{air\ change}\right)(12°F)$$
$$= 64{,}800\ Btu/hr$$

The total heat loss from conduction and infiltration from air exchange is

$$\begin{aligned}(q_{70°F} - q_{58°F}) \\ + (q_{infiltration,70°F} - q_{infiltration,58°F})\end{aligned} = 69{,}000 \frac{Btu}{hr}$$

$$+ 64{,}800 \frac{Btu}{hr}$$

$$= 133{,}800\ Btu/hr$$

The number of unoccupied hours per heating season is

$$\left(23 \frac{wk}{yr}\right)\left(\begin{array}{c}\left(14 \frac{hr}{day}\right)\left(5 \frac{day}{wk}\right) \\ + \left(24 \frac{hr}{day}\right)\left(2 \frac{day}{wk}\right)\end{array}\right) = 2714\ hr/yr$$

The energy savings is

$$\frac{\left(2714 \frac{hr}{yr}\right)\left(133{,}800 \frac{Btu}{hr}\right)}{(0.77)\left(1 \times 10^5 \frac{Btu}{therm}\right)} = 4716\ therm/yr$$

The savings is

$$\left(0.30 \frac{\$}{therm}\right)\left(4716 \frac{therm}{yr}\right) = \$1414.80 \quad (\$1400)$$

The answer is (C).

50. Find the annual number of operating hours for the light bulbs.

$$\left(6 \frac{hr}{day}\right)\left(7 \frac{day}{wk}\right)\left(32 \frac{wk}{yr}\right) = 1344\ hr/yr$$

The annual energy savings from LED light bulbs is

$$energy = power \times time$$

$$= \left(\begin{array}{c}(43\ bulbs)\left(\frac{100\ W - 14\ W}{bulb}\right) \\ + (14\ bulbs)\left(\frac{60\ W - 8.5\ W}{bulb}\right) \\ + (10\ bulbs)\left(\frac{40\ W - 4.5\ W}{bulb}\right)\end{array}\right)$$

$$\times \left(1344 \frac{h}{yr}\right)\left(10^{-3} \frac{kW}{W}\right)$$

$$= 6{,}416\ kWh/yr$$

The annual savings from switching to LED bulbs is

$$\left(6416 \ \frac{\text{kWh}}{\text{yr}}\right)\left(0.9 \ \frac{\$}{\text{kWh}}\right) = \$577.46$$

Find the sensible heat lost from switching to LED bulbs.

Electric Lighting

$q_d = 3.412 \, W F_{ul} F_{sa}$

$$= \left(3.412 \ \frac{\text{Btu}}{\text{W-hr}}\right) \begin{pmatrix} (43 \text{ bulbs})\left(\dfrac{100 \text{ W} - 14 \text{ W}}{\text{bulbs}}\right)(0.97) \\ +(14 \text{ bulbs})\left(\dfrac{60 \text{ W} - 8.5 \text{ W}}{\text{bulbs}}\right)(0.95) \\ +(10 \text{ bulbs})\left(\dfrac{40 \text{ W} - 4.5 \text{ W}}{\text{bulbs}}\right)(0.90) \end{pmatrix}$$

$$\times \left(1344 \ \frac{\text{hr}}{\text{yr}}\right)$$

$= 21{,}055{,}416 \text{ Btu/yr}$

Find the additional therms needed from the furnace to make up for the lost heat.

$$\frac{21{,}055{,}416 \ \dfrac{\text{Btu}}{\text{yr}}}{(0.8)\left(1 \times 10^5 \ \dfrac{\text{Btu}}{\text{therm}}\right)} = 263.19 \text{ therm}$$

The additional cost of natural gas to make up for the lost heat is

$$(263.19 \text{ therm})\left(\frac{\$1.20}{\text{therm}}\right) = \$315.83$$

Subtract the additional cost of natural gas from the electrical savings to find the total savings.

$$\$577.46 - \$315.83 = \$261.63 \quad (\$260)$$

The answer is (B).

51. The municipality will be performing the analysis to ensure the property owner is in compliance with the municipal noise ordinances. The item being analyzed is a chiller and it is located outdoors. From a table of comparison of sound rating methods, the most likely method to be used is A-weighted decibels (dBA). [Comparison of Sound Rating Methods]

The answer is (D).

52. As water progresses down the sprinkler line it will undergo a pressure drop. The last sprinkler in the sprinkler line will have the lowest pressure, which is required to be a minimum of 10 psig. Start with the equation for flow in an orifice.

Orifices

$$Q = CA_0 \sqrt{2g\left(\frac{p_1}{\gamma} + z_1 - \frac{p_2}{\gamma} - z_2\right)}$$

There is no change in height described in the problem statement, so $z_1 = z_2 = 0$. If point 1 is inside the pipe and point 2 is at the point of discharge, $p_1 = 10$ psig and $p_2 = 0$ psig. The flow through the last sprinkler head is

$$Q_1 = CA_0 \sqrt{2g\left(\frac{p_1}{\gamma}\right)} = CA_0 \sqrt{2\left(\frac{p_1}{\rho}\right)}$$

$$= \left(\frac{(0.75)\pi(0.5 \text{ in})^2}{(4)\left(144 \ \dfrac{\text{in}^2}{\text{ft}^2}\right)}\right)$$

$$\times \sqrt{\frac{(2)\left(10 \ \dfrac{\text{lbf}}{\text{in}^2}\right)\left(144 \ \dfrac{\text{in}^2}{\text{ft}^2}\right)\left(32.2 \ \dfrac{\text{lbm-ft}}{\text{lbf-sec}^2}\right)}{\left(62.4 \ \dfrac{\text{lbm}}{\text{ft}^3}\right)}}$$

$= 0.0394 \text{ ft}^3/\text{sec}$

Converting to gallons per minute gives

$$Q_1 = \left(0.0394 \ \frac{\text{ft}^3}{\text{sec}}\right)\left(60 \ \frac{\text{sec}}{\text{min}}\right)\left(7.46 \ \frac{\text{gal}}{\text{ft}^3}\right) = 17.65 \text{ gal/min}$$

Recognize that the flow to the second-to-last head (point 3) will be greater than the flow through the last head (point 1) since the last head is at a minimum pressure. In other words, answer options A and B can be neglected. The pressure loss in a 1 in line over 10 ft is negligible. The flow at the second-to-last head is

$$Q_3 = CA_0 \sqrt{2g\left(\frac{p_3}{\gamma}\right)}$$

But the pressure is different due to head loss in the line between the two sprinklers. The pressure in the second-to-last head can be found using the Bernoulli equation.

Bernoulli Equation

$$\frac{p_3}{\rho g} + z_3 + \frac{v_3^2}{2g} = \frac{p_1}{\rho g} + z_1 + \frac{v_1^2}{2g} + h_f$$

The heads are at the same elevation, and the dynamic head is negligible. The equation for the pressure in the second-to-last head is

$$\frac{p_3}{\rho g} = \frac{p_1}{\rho g} + h_f$$
$$p_3 = p_1 + \rho g h_f$$

The head loss per foot of pipe, $h_{f,\text{ft/ft}}$, can be found using the Hazen-Williams equation.

Pressure Drop of Water Flowing in Circular Pipe (Hazen-Williams)

$$\begin{aligned} h_{f,\text{ft/ft}} &= \frac{10.44 Q^{1.85}}{C^{1.85} D^{4.8655}} \\ &= \frac{(10.44)\left(17.65\ \frac{\text{gal}}{\text{min}}\right)^{1.85}}{(100)^{1.85}(1\ \text{in})^{4.87}} \\ &= 0.4219\ \text{ft/ft} \end{aligned}$$

The sprinklers are located 10 ft apart. The head loss in feet for 10 ft of pipe is

$$\begin{aligned} h_f &= \left(0.4219\ \frac{\text{ft}}{\text{ft}}\right)(10\ \text{ft}) \\ &= 4.219\ \text{ft} \end{aligned}$$

The pressure in the second-to-last sprinkler head is

$$\begin{aligned} p_3 &= p_1 + \rho g h_f \\ &= 10\ \frac{\text{lbf}}{\text{in}^2} + \frac{\left(62.4\ \frac{\text{lbm}}{\text{ft}^3}\right)\left(32.2\ \frac{\text{ft}}{\text{sec}^2}\right)(4.219\ \text{ft})}{\left(32.2\ \frac{\text{lbm-ft}}{\text{lbf-sec}^2}\right)\left(12\ \frac{\text{in}}{\text{ft}}\right)^2} \\ &= 11.82\ \text{lbf}/\text{in}^2 \end{aligned}$$

Solving the previous equation for the flow rate through the second-to-last sprinkler head gives

$$\begin{aligned} Q_3 &= CA_0\sqrt{2g_c\left(\frac{p_3}{\gamma}\right)} \\ &= \frac{(0.75)\pi(0.5\ \text{in})^2}{(4)\left(12\ \frac{\text{in}}{\text{ft}}\right)^2} \\ &\quad \times \sqrt{\frac{(2)\left(32.2\ \frac{\text{lbm-ft}}{\text{lbf-sec}^2}\right)\left(11.82\ \frac{\text{lbf}}{\text{in}^2}\right)\left(12\ \frac{\text{in}}{\text{ft}}\right)^2}{62.4\ \frac{\text{lbm}}{\text{ft}^3}}} \\ &= 0.042\ \text{ft}^3/\text{sec} \end{aligned}$$

Converting to gallons per minute gives

$$\begin{aligned} Q_3 &= \left(0.042\ \frac{\text{ft}^3}{\text{sec}}\right)\left(60\ \frac{\text{sec}}{\text{min}}\right)\left(7.46\ \frac{\text{gal}}{\text{ft}^3}\right) \\ &= 19.19\ \text{gal/min}\ \ (19\ \text{gpm}) \end{aligned}$$

The answer is (C).

53. Use the Darcy-Weisbach equation to find the head loss. The inner diameter of the pipe in feet is

$$D = \frac{6\ \text{in}}{12\ \frac{\text{in}}{\text{ft}}} = 0.5\ \text{ft}$$

Find the Reynolds number. [Properties of Water (I-P Units)]

Reynolds Number

$$\begin{aligned} \text{Re} &= \frac{\text{v} D \rho}{\mu g_c} \\ &= \frac{\left(5\ \frac{\text{ft}}{\text{sec}}\right)(0.5\ \text{ft})\left(62.4\ \frac{\text{lbm}}{\text{ft}^3}\right)}{\left(2.359 \times 10^{-5}\ \frac{\text{lbf-sec}}{\text{ft}^2}\right)\left(32.2\ \frac{\text{lbm-ft}}{\text{lbf-sec}^2}\right)} \\ &= 205{,}400 \end{aligned}$$

The relative roughness is

$$\frac{\varepsilon}{D} = \frac{0.0008\ \text{ft}}{0.5\ \text{ft}} = 0.0016$$

From these values and a Moody diagram, the Darcy friction factor is 0.023. [Moody Diagram (Stanton Diagram)]

From the Darcy-Weisbach equation, the head loss is

Head Loss Due to Flow: Darcy-Weisbach Equation

$$h_f = f\frac{L}{D}\frac{\text{v}^2}{2g} = \frac{(0.023)\left(\frac{2000\ \text{ft}}{0.5\ \text{ft}}\right)\left(5\ \frac{\text{ft}}{\text{sec}}\right)^2}{(2)\left(32.2\ \frac{\text{ft}}{\text{sec}^2}\right)} = 35.7\ \text{ft}\ \ (35\ \text{ft})$$

The answer is (D).

54. Find the total thermal resistance. [Thermal Resistance (R)]

$$R_{total} = \frac{1}{h_o A_o} + \frac{1}{2\pi k L}\ln\frac{r_o}{r_i} + \frac{1}{h_i A_o}$$

$$= \frac{(1)\left(12\,\frac{\text{in}}{\text{ft}}\right)}{\left(2.0\,\frac{\text{Btu}}{\text{hr-ft}^2\text{-}°\text{F}}\right)\pi(1.5\,\text{in})(120\,\text{ft})}$$

$$+ \frac{1}{2\pi\left(29\,\frac{\text{Btu}}{\text{hr-ft-}°\text{F}}\right)(120\,\text{ft})}\ln\left(\frac{1.5\,\text{in}}{1.37\,\text{in}}\right)$$

$$+ \frac{(1)\left(12\,\frac{\text{in}}{\text{ft}}\right)}{\left(1500\,\frac{\text{Btu}}{\text{hr-ft}^2\text{-}°\text{F}}\right)\pi(1.37\,\text{in})(120\,\text{ft})}$$

$$= 0.0106\,\text{hr-}°\text{F/Btu}$$

The rate of heat transfer from the pipe to the ambient air is

Thermal Resistance (R)

$$q = \frac{\Delta T}{R_{total}}$$

The saturation temperature of steam at 1 atm is 212°F. The heat loss over the entire length of uninsulated pipe is

$$q = \frac{212°\text{F} - 60°\text{F}}{0.0106\,\frac{\text{hr-}°\text{F}}{\text{Btu}}} = 14{,}300\,\text{Btu/hr} \quad (14{,}000\,\text{Btu/hr})$$

The answer is (A).

55. Using an efficiency equation, find the annual heat energy consumption in therms.

Basic Cycles

$$\eta = \frac{W}{Q_H} = \frac{Q_{out}}{Q_{in}}$$

The standard furnace consumes 1500 therms, so

$$Q_{out} = Q_{in}\eta = \left(1500\,\frac{\text{therms}}{\text{yr}}\right)(0.78) = 1170\,\text{therms/yr}$$

The high-efficiency furnace will have the same Q_{out} but a different Q_{in}.

$$Q_{in} = \frac{1170\,\frac{\text{therms}}{\text{yr}}}{0.92} = 1272\,\text{therms/yr}$$

The annual energy savings are

$$1500\,\frac{\text{therms}}{\text{yr}} - 1272\,\frac{\text{therms}}{\text{yr}} = 228\,\text{therms/yr}$$

The annual cost savings are

$$\left(228\,\frac{\text{therms}}{\text{yr}}\right)\left(0.85\,\frac{\$}{\text{therm}}\right) = \$194/\text{yr}$$

Simple payback on investment is one of the most common indexes of economic merit used when there is a need for quick decisions. The additional investment cost is divided by the annual savings attributed solely to that investment. Payback, therefore, equals incremental cost divided by annual savings.

$$t = \frac{C}{A}$$

The simple payback period is

$$t = \frac{\$800}{\frac{\$194}{\text{yr}}} = 4.1\,\text{yr} \quad (4\,\text{yr})$$

The answer is (B).

56. Calculate the heat flux from the building structure.

Composite Plane Wall

$$q'' = UA(T_{structure} - T_{environment})$$

$$\frac{q''}{A} = U(T_{structure} - T_{environment})$$

$$= \left(0.9\,\frac{\text{Btu}}{\text{hr-ft}^2\text{-}°\text{F}}\right)(20°\text{F})$$

$$= 18\,\text{Btu/hr-ft}^2$$

After insulation is added,

$$\frac{q''}{A} = U(T_{structure} - T_{environment}) = \left(0.6\,\frac{\text{Btu}}{\text{hr-ft}^2\text{-}°\text{F}}\right)(20°\text{F})$$

$$= 12\,\text{Btu/hr-ft}^2$$

Determine the annual energy savings.

$$\left(18 \, \frac{\text{Btu}}{\text{hr-ft}^2} - 12 \, \frac{\text{Btu}}{\text{hr-ft}^2}\right) \\ \times \left(24 \, \frac{\text{hr}}{\text{day}}\right)\left(180 \, \frac{\text{day}}{\text{yr}}\right) \\ = 25{,}920 \, \text{Btu/yr-ft}^2$$

Determine the annual cost savings.

$$\left(25{,}920 \, \frac{\text{Btu}}{\text{yr-ft}^2}\right)\left(\frac{\$6}{1{,}000{,}000 \, \text{Btu}}\right) = \$0.156/\text{yr-ft}^2$$

The payback period of the insulation is

$$\frac{0.75 \, \frac{\$}{\text{ft}^2}}{0.156 \, \frac{\$}{\text{yr-ft}^2}} = 4.82 \, \text{yr} \quad (5 \, \text{yr})$$

The answer is (D).

57. The water vapor is at the same dry-bulb temperature as the air, 80°F. From a psychrometric chart or saturated steam tables, the water vapor saturation pressure at a dry-bulb temperature of 80°F is 0.5073 psia. [Properties of Saturated Water and Steam (Pressure) - I-P Units]

The relative humidity is the ratio of the vapor pressure to the saturation pressure of water at a given temperature. Solving for the partial pressure of the water vapor gives

Psychrometric Properties

$$\phi = \frac{p_w}{p_{w,\text{sat}}}\bigg|_{\text{tp}}$$

$$p_w = \phi p_{w,\text{sat}} = (0.40)\left(0.5073 \, \frac{\text{lbf}}{\text{in}^2}\right) = 0.203 \, \text{lbf/in}^2 \quad (0.20 \, \text{psia})$$

The partial pressure of the water vapor is independent of the total pressure or the pressure of the other component, air.

The answer is (A).

58. Find the total conduction resistance. [Thermal Resistance (R)]

$$R_{\text{total}} = \frac{1}{h_i A_i} + \frac{1}{2\pi k L} \ln \frac{r_o}{r_i}$$

Since this is thin-walled pipe, $r_o \approx r_i$ and $\ln(1) = 0$.

$$R_{\text{total}} = \frac{1}{h_i A_i} = \frac{(1)\left(12 \, \frac{\text{in}}{\text{ft}}\right)}{\left(5 \, \frac{\text{W}}{\text{ft}^2\text{-}°\text{F}}\right)\pi\left(\frac{5}{8} \, \text{in}\right)L} \\ = \frac{1.22 \, \frac{°\text{F-ft}}{\text{W}}}{L}$$

The linear heat rate given in the problem statement is

$$q_{\text{per ft}} = \frac{q}{L} = 25 \, \text{W/ft}$$

Use the equation for the rate of heat transfer, and solve for the temperature of the ambient air.

Thermal Resistance (R)

$$q = \frac{\Delta T}{R_{\text{total}}}$$

$$T_\infty - T_i = qR_{\text{total}}$$

$$T_\infty = qR_{\text{total}} + T_i$$

$$= (25 \, \text{W})\left(\frac{1.22 \, \frac{°\text{F-ft}}{\text{W}}}{L \, (\text{in feet})}\right) + (-5°\text{F})$$

$$= \left(25 \, \frac{\text{W}}{\text{ft}}\right)\left(1.22 \, \frac{°\text{F-ft}}{\text{W}}\right) + (-5°\text{F})$$

$$= 25.6°\text{F} \quad (26°\text{F})$$

The answer is (A).

59. At 80°F and a relative humidity of zero (since the air is dry), the specific humidity of the entering air is zero. From saturated air tables or a psychrometric chart, at 110°F and a relative humidity of 100%, the specific humidity of the exiting air is 0.0610 lbm/lbm. [ASHRAE Psychrometric Chart No. 1 - Normal Temperature at Sea Level]

The channel is insulated, so it can be treated as an adiabatic, steady-flow system with two inlets and one outlet. From conservation of energy, the amount of moisture picked up by the air equals the amount of moisture evaporated from the water.

Adiabatic Mixing of Water Injected Into Moist Air (Evaporative Cooling)

$$\dot{m}_{\text{da}} W_1 + \dot{m}_w = \dot{m}_{\text{da}} W_2$$

$$W_1 = 0$$

The mass flow rate of the water is

$$\dot{m}_w = \left(30 \ \frac{\text{lbm}}{\text{sec}}\right)\left(0.061 \ \frac{\text{lbm}}{\text{lbm}}\right) = 1.83 \ \text{lbm/sec}$$

The density of water is 8.34 lbm/gal. The rate that moisture is picked up by the air in gallons per minute is

$$\frac{\left(1.83 \ \frac{\text{lbm}}{\text{sec}}\right)\left(60 \ \frac{\text{sec}}{\text{min}}\right)}{8.34 \ \frac{\text{lbm}}{\text{gal}}} = 13.16 \ \text{gal/min} \quad (13 \ \text{gpm})$$

The answer is (D).

60. The total minor losses are the sum of the loss contributions of each of the fittings. There are two 90° elbows and one 45° elbow, so

$$K_{\text{total}} = (2)(0.9) + 0.42 = 2.22$$

From a table of properties for schedule-40 steel pipe, the inside diameter of the pipe is 0.824 in. [Schedule 40 Steel Pipe]

From the continuity equation, the velocity through the pipe is

Continuity Equation

$$Q = Av$$

$$v = \frac{Q}{A} = \frac{\left(48 \ \frac{\text{gal}}{\text{min}}\right)\left(0.134 \ \frac{\text{ft}^2}{\text{gal}}\right)\left(12 \ \frac{\text{in}}{\text{ft}}\right)^2}{\left(\frac{\pi}{4}\right)(0.824 \ \text{in})^2\left(60 \ \frac{\text{sec}}{\text{min}}\right)} = 28.95 \ \text{ft/sec}$$

The sum of the minor losses for the system is

Valve and Fittings Losses

$$\Delta h = K\frac{v^2}{2g} = \frac{(2.22)\left(28.95 \ \frac{\text{ft}}{\text{sec}}\right)^2}{(2)\left(32.2 \ \frac{\text{ft}}{\text{sec}^2}\right)}$$

$$= 28.9 \ \text{ft} \quad (29 \ \text{ft})$$

The answer is (A).

61. From the psychrometric chart, air at 40°F and 20% relative humidity has a specific humidity of 0.001 lbm/lbm and a specific volume of 12.62 ft³/lbm. Relevant information from the chart is shown. [ASHRAE Psychrometric Chart No. 1 - Normal Temperature at Sea Level] [Moist-Air Cooling and Dehumidification]

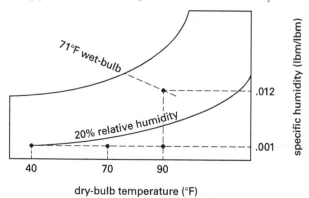

The addition of sensible heat to 70°F does not affect the specific humidity. From the psychrometric chart, the air at the final condition of 90°F dry-bulb temperature and 71°F wet-bulb temperature has a specific humidity of 0.012 lbm/lbm.

The mass flow rate at which hot steam is needed is

$$\dot{m}_{\text{steam}} = \dot{m}_{\text{air}}(W_2 - W_1)$$

$$= \left(\frac{6000 \ \frac{\text{ft}^3}{\text{min}}}{12.62 \ \frac{\text{ft}^3}{\text{lbm}}}\right)\left(0.012 \ \frac{\text{lbm}}{\text{lbm}} - 0.001 \ \frac{\text{lbm}}{\text{lbm}}\right)$$

$$= 5.23 \ \text{lbm/min}$$

From steam tables, the specific volume of steam at atmospheric pressure is 26.78 ft³/lbm. [Properties of Saturated Water and Steam (Pressure) - I-P Units]

The volumetric flow rate is

$$\dot{m}_{\text{steam}} = \rho_{\text{steam}}Q_{\text{steam}}$$

$$Q_{\text{steam}} = \frac{\dot{m}_{\text{steam}}}{\rho_{\text{steam}}}$$

$$= \dot{m}_{\text{steam}}v_{\text{steam}}$$

$$= \frac{\left(5.23 \ \frac{\text{lbm}}{\text{min}}\right)\left(26.78 \ \frac{\text{ft}^3}{\text{lbm}}\right)}{60 \ \frac{\text{sec}}{\text{min}}}$$

$$= 2.33 \ \text{ft}^3/\text{sec} \quad (2.3 \ \text{ft}^3/\text{sec})$$

The answer is (C).

62. At maximum design load, the pump speed is 3450 rpm. The affinity laws indicate the following.

Pump Affinity Laws

$$Q_2 = Q_1 \left(\frac{N_2}{N_1}\right)$$

$$N_2 = N_1 \left(\frac{Q_2}{Q_1}\right)$$

$$= (3450 \text{ rpm})\left(\frac{1}{2}\right)$$

$$= 1725 \text{ rpm}$$

The affinity laws indicate that when flow is reduced by 50%, speed is reduced by 50%, which in this case would be 1725 rpm. However, the pump is part of a larger application, and the system curve for this application does not start at zero head.

Half the design flow is 500 gpm/2 = 250 gpm. According to the graph, 250 gpm occurs with a speed nearest to 2587 rpm (2600 rpm).

The answer is (C).

63. The volume of the tank is

$$V = \pi r^2 H = \left(\frac{\pi}{4}\right)(7.25 \text{ in})^2 \left(\frac{30 \text{ in}}{\left(12 \frac{\text{in}}{\text{ft}}\right)^3}\right)$$

$$= 0.717 \text{ ft}^3$$

The initial mass of air in the tank is

$$(80 \text{ ft}^3)\left(0.075 \frac{\text{lbm}}{\text{ft}^3}\right) = 6 \text{ lbm}$$

The mass of air when there is 700 psig is

Ideal Gas

$$pV = mRT$$

$$m = \frac{pV}{RT}$$

$$= \frac{\left(700 \frac{\text{lbf}}{\text{in}^2} + 14.7 \frac{\text{lbf}}{\text{in}^2}\right)\left(12 \frac{\text{in}}{\text{ft}}\right)^2 (0.717 \text{ ft}^3)}{\left(53.3 \frac{\text{ft-lbf}}{\text{lbm-°R}}\right)(70°\text{F} + 460°)}$$

$$= 2.61 \text{ lbm}$$

The diver can breathe 3.39 lbm of air before the tank reaches low pressure and is considered empty. To find the rate of air the diver needs to perform the work, use the maximum recommended value for the level of exertion. From an ASHRAE table of oxygen consumption at different activity levels, moderate work consumes 1–2 ft³/hr of air. [Heart Rate and Oxygen Consumption at Different Activity Levels]

The mass of air the diver consumes per hour is

$$\left(2 \frac{\text{ft}^3}{\text{hr}}\right)\left(0.075 \frac{\text{lbm}}{\text{ft}^3}\right) = 0.15 \text{ lbm/hr}$$

The number of hours of air in the tank for performing moderate work is

$$\frac{6 \text{ lbm} - 2.61 \text{ lbm}}{0.15 \frac{\text{lbm}}{\text{hr}}} = 22.6 \text{ hr} \quad (23 \text{ hr})$$

The answer is (C).

64. The density of the air at the inlet is

$$\rho = \frac{p}{RT} = \frac{\left(92 \frac{\text{lbf}}{\text{in}^2}\right)\left(12 \frac{\text{in}}{\text{ft}}\right)^2}{\left(53.3 \frac{\text{ft-lbf}}{\text{lbm-°R}}\right)(25°\text{F} + 460°)} = 0.512 \text{ lbm/ft}^3$$

The compressor power in adiabatic compression is

Compressors

$$P_{\text{comp}} = \frac{\dot{m} p_i k}{(k-1)\rho_i \eta_c}\left[\left(\frac{p_e}{p_i}\right)^{1-\frac{1}{k}} - 1\right]$$

Divide the compressor power by the mass flow rate to find the specific work of the compressor.

$$w = \frac{P_{\text{comp}}}{\dot{m}} = \frac{p_i k}{(k-1)\rho_i \eta_c}\left[\left(\frac{p_e}{p_i}\right)^{1-\frac{1}{k}} - 1\right]$$

$$= \frac{\left(92 \frac{\text{lbf}}{\text{in}^2}\right)\left(12 \frac{\text{in}}{\text{ft}}\right)^2 (1.4)}{(1.4-1)\left(0.512 \frac{\text{lbm}}{\text{ft}^3}\right)(0.84)}\left[\left(\frac{260 \frac{\text{lbf}}{\text{in}^2}}{92 \frac{\text{lbf}}{\text{in}^2}}\right)^{1-\frac{1}{1.4}} - 1\right]$$

$$= 37{,}258 \text{ ft-lbf/lbm} \quad (37{,}000 \text{ ft-lbf/lbm})$$

The answer is (D).

65. To compare two alternatives using payback analysis, determine the annual energy cost for each option.

$$C_{\text{annual}} = E_{\text{kW-hr/ton}} C_{\$/\text{kW-hr}} n_{\text{hr/yr}}$$

$$C_{\text{annual},1} = \left(0.8 \, \frac{\text{kW}}{\text{ton}}\right)\left(0.10 \, \frac{\$}{\text{kW-hr}}\right)\left(2000 \, \frac{\text{hr}}{\text{yr}}\right)$$
$$= \$160/\text{ton-yr}$$

$$C_{\text{annual},2} = \left(0.7 \, \frac{\text{kW}}{\text{ton}}\right)\left(0.10 \, \frac{\$}{\text{kW-hr}}\right)\left(2000 \, \frac{\text{hr}}{\text{yr}}\right)$$
$$= \$140/\text{ton-yr}$$

Option 2 has the higher cost efficiency. The payback period is given by the difference in installed costs divided by the difference in annual costs.

$$\text{payback period} = \frac{\Delta C_{\text{install}}}{\Delta C_{\text{annual}}}$$
$$= \frac{C_{\text{install},1} - C_{\text{install},2}}{C_{\text{annual},1} - C_{\text{annual},2}}$$
$$= \frac{350 \, \frac{\$}{\text{ton}} - 300 \, \frac{\$}{\text{ton}}}{160 \, \frac{\$}{\text{ton-yr}} - 140 \, \frac{\$}{\text{ton-yr}}}$$
$$= 2.5 \, \text{yr}$$

The answer is (A).

66. (*Fill in the blank*) Heat is transferred from the exhaust air to the makeup air. From the psychrometric chart, the specific enthalpy of the makeup air before conditioning is 9.3 Btu/lbm. Relevant information from the chart is shown. [ASHRAE Psychrometric Chart No. 1 - Normal Temperature at Sea Level]

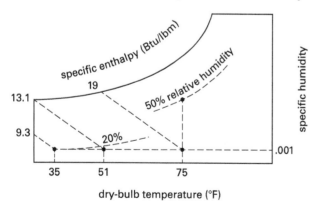

The exhaust air undergoes a latent heat change as it transfers heat to the makeup air. After conditioning, the exhaust air has the same humidity ratio as the makeup air. On the psychrometric chart, find the states of the exhaust air and makeup air before conditioning. Draw a horizontal line to the right from the makeup air, and draw a vertical line down from the exhaust air. The point where these two lines intersect is the state of the exhaust air after conditioning. This point has a specific enthalpy of approximately 19 Btu/lbm. The sensible heat difference between the exhaust air after conditioning and the makeup air before conditioning is

$$19 \, \frac{\text{Btu}}{\text{lbm}} - 9.3 \, \frac{\text{Btu}}{\text{lbm}} = 9.7 \, \text{Btu/lbm}$$

The sensible heat recovered is 40% of this, or

$$(0.4)\left(9.7 \, \frac{\text{Btu}}{\text{lbm}}\right) = 3.88 \, \text{Btu/lbm} \quad (4 \, \text{Btu/lbm})$$

Alternative Solution

The problem can also be solved more quickly by recognizing that, because the exhaust and makeup airflow rates are equal, a 40% sensible heat transfer results in a 40% dry-bulb temperature rise. [**Heat-Recovery Ventilator (HRV)—Sensible Energy Recovery**]

The dry-bulb temperature of the makeup air rises from 35°F to

$$(35°F)(1.4) = 49°F$$

From the psychrometric chart, the conditioned makeup air at 49°F has a specific enthalpy of approximately 13.0 Btu/lbm. The increase in specific enthalpy is

$$13.0 \, \frac{\text{Btu}}{\text{lbm}} - 9.3 \, \frac{\text{Btu}}{\text{lbm}} = 3.8 \, \text{Btu/lbm} \quad (4 \, \text{Btu/lbm})$$

The answer is (B).

67. The three-way valve shown is a mixing valve, which is commonly used to mix heating water to a desired temperature. The flow rate through the pump and the pressure drop across the pump is the same (or nearly the same) no matter how the valve operates. If maintenance on the boiler is necessary, flow through the entire loop would be stopped and the boiler would be isolated with gate valves near the boiler vessel.

The answer is (C).

68. Find the needed diameter of the new duct by using the equation for the relationship between rectangular and round ducts.

Rectangular Ducts

$$D_c = \frac{1.30(ab)^{0.625}}{(a+b)^{0.250}}$$

$$= \left(\frac{(1.30)\big((2\text{ ft})(5\text{ ft})\big)^{0.625}}{(2\text{ ft}+5\text{ ft})^{0.250}}\right)\left(12\ \frac{\text{in}}{\text{ft}}\right)$$

$$= 40.44\text{ in}\quad(40\text{ in})$$

The answer is (B).

69. Based on ASHRAE Standard 62.1, the minimum distance is 25 ft from the place where vehicle exhaust is likely to be located. [Air Intake Minimum Separation Distance, Based on ANSI/ASHRAE Standard 62.1-2007]

The answer is (B).

70. The only reason for pressure drop in the ductwork is friction. From the Darcy-Weisbach equation,

Head Loss Due to Flow: Darcy-Weisbach Equation

$$h_f = f\left(\frac{L}{D}\right)\left(\frac{\text{v}^2}{2g}\right)$$

The pressure available to overcome friction will determine how fast the air will flow. For both the straight branch and the take-off branch, this pressure is equal to the pressure at the branch minus what is needed to drive the diffusers. Each duct has the same head loss.

$$h_f = 0.10\text{ in wg} - 0.03\text{ in wg} = 0.07\text{ in wg}$$

The equation can be rewritten to include flow rates as

$$h_f = f\left(\frac{L}{D}\right)\left(\frac{\text{v}^2}{2g}\right)$$

$$= \frac{fL}{2Dg}\left(\frac{Q}{A}\right)^2$$

$$= \frac{fL}{2Dg}\left(\frac{Q}{\frac{\pi D^2}{4}}\right)^2$$

$$= \frac{8fLQ^2}{\pi^2 D^5 g}$$

The lengths and diameters of the ducting are the only differences between the two branches. The simplified head loss equation for the two branches is

$$\frac{L_{20}}{D_{20}^5} = \frac{L_{35}}{D_{35}^5}$$

$$\frac{D_{20}^5}{D_{35}^5} = \frac{20\text{ ft}}{35\text{ ft}}$$

$$\frac{D_{20}}{D_{35}} = \sqrt[5]{\frac{20\text{ ft}}{35\text{ ft}}}$$

$$= 0.894$$

The diameter of the take-off branch is

$$\frac{D_{20}}{0.894} = \frac{11\text{ in}}{0.894}$$

$$= 12.3\text{ in}\quad(13\text{ in})$$

The answer is (C).

71. The refrigerating effect is the enthalpy difference between the condition of the refrigerant when entering the evaporator and its condition when leaving. Use a pressure-enthalpy diagram for refrigerant-134a. [Pressure Versus Enthalpy Curves for Refrigerant 134a]

When the refrigerant enters the evaporator at 34.7 psia and 15% quality, its enthalpy, h, is

$$h_{\text{in}} = 33\text{ Btu/lbm}$$

When the refrigerant leaves the evaporator at 29.7 psia and 100% quality, its enthalpy is

$$h_{\text{out}} = 105\text{ Btu/lbm}$$

The refrigerating capacity of the system is

$$\dot{Q} = \dot{m}(h_{\text{out}} - h_{\text{in}})$$

$$= \frac{\left(10\ \frac{\text{lbm}}{\text{min}}\right)\left(105\ \frac{\text{Btu}}{\text{lbm}} - 33\ \frac{\text{Btu}}{\text{lbm}}\right)\left(60\ \frac{\text{min}}{\text{hr}}\right)}{12{,}000\ \frac{\text{Btu}}{\text{ton-hr}}}$$

$$= 3.6\text{ tons}\quad(4\text{ tons})$$

The answer is (A).

72. Waste heat refers to energy that is generated but not put to practical use. While some waste heat losses from industrial processes are inevitable, losses can be reduced by improving equipment efficiency and by installing waste heat recovery technologies. Waste heat recovery entails capturing and reusing the waste heat from industrial processes or the heat lost from air conditioning systems.

Waste heat can take the forms of both sensible and latent heat. Latent heat is often the greater part, as in combustion gas exhaust.

However, recovering latent waste heat is not as simple as recovering sensible waste heat (high-temperature streams). To recover latent waste heat, the transfer material must come in direct contact with the exhausting stream and be able to absorb and evaporate liquids.

Heat recovery ventilators are for sensible energy recovery. Heat pipes and heat pumps are indirect methods and do not come into direct contact with the exhausting stream.

Energy recovery ventilators come in direct contact with the exhausting stream and have the capability to recover both sensible and latent heat. [Energy-Recovery Ventilator (ERV)]

The answer is (A).

73. Find the output cooling energy.

$$\text{output} = (40 \text{ tons})\left(12{,}000 \; \frac{\text{Btu}}{\text{hr-ton}}\right)$$
$$= 480{,}000 \; \text{Btu/hr}$$

To find the energy efficiency ratio (EER), divide the output cooling energy in British thermal units per hour by the electrical load in watts.

Efficiency

$$\text{EER} = \frac{\text{output cooling energy (Btu/hr)}}{\text{input electrical energy (W)}}$$
$$= \frac{480{,}000 \; \frac{\text{Btu}}{\text{hr}}}{55{,}000 \; \text{W}}$$
$$= 8.727 \quad (8.7)$$

The answer is (A).

74. Use the pressure-enthalpy (p-h) diagram for refrigerant-134a and the single-stage refrigeration cycle to determine the mass flow rate. The illustration shows the p-h diagram for this cycle, including the 20°F superheat and 10°F subcooling. [Pressure Versus Enthalpy Curves for Refrigerant 134a]

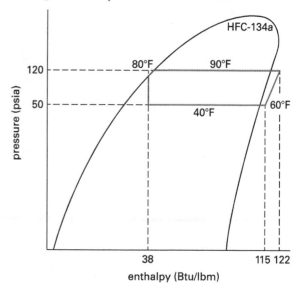

Conditions at the evaporator are

$$h_{\text{in}} = 38 \; \text{Btu/lbm}$$
$$h_{\text{out}} = 115 \; \text{Btu/lbm}$$

The capacity in British thermal units per hour is

$$q = (3 \text{ tons})\left(12{,}000 \; \frac{\text{Btu}}{\text{hr-ton}}\right)$$
$$= 36{,}000 \; \text{Btu/hr}$$

The mass flow rate is

$$\dot{m} = \frac{q}{h_{\text{out}} - h_{\text{in}}} = \frac{\left(36{,}000 \; \frac{\text{Btu}}{\text{hr}}\right)}{\left(115 \; \frac{\text{Btu}}{\text{lbm}} - 38 \; \frac{\text{Btu}}{\text{lbm}}\right)\left(3600 \; \frac{\text{sec}}{\text{hr}}\right)}$$
$$= 0.13070 \; \text{lbm/sec}$$

The conditions at the compressor are

$$h_{\text{in}} = 115 \; \text{Btu/lbm}$$
$$h_{\text{out}} = 122 \; \text{Btu/lbm}$$

The power needed for the compressor is

Compressors

$$P = \dot{m}(h_{in} - h_{out})$$

$$= \frac{\left(0.130 \ \frac{\text{lbm}}{\text{sec}}\right)\left(122 \ \frac{\text{Btu}}{\text{lbm}} - 115 \ \frac{\text{Btu}}{\text{lbm}}\right)\left(778 \ \frac{\text{ft-lbf}}{\text{Btu}}\right)}{550 \ \frac{\text{ft-lbf}}{\text{hp-sec}}}$$

$$= 1.29 \text{ hp} \quad (1.3 \text{ hp})$$

The answer is (B).

75. The mass flow rate of water is related to the mass flow rate of the air and the entering and exiting humidity ratios.

$$\dot{m}_w = \dot{m}_{da}(W_1 - W_2) = \frac{Q}{v}(W_1 - W_2)$$

The required mass flow rate of dry air for a health club can be found from a table of air and ventilation requirements for various facilities. A health club requires a combined outdoor air rate of 22 ft³/person and an area outdoor air rate of 0.06 ft³/ft². [Minimum Ventilation Rates in the Breathing Zone, Based on ANSI/ASHRAE Standard 62.1-2007]

The total flow rate for the building is

$$Q = (8000 \text{ ft}^2)\left(\left(\frac{40 \text{ persons}}{1000 \text{ ft}^2}\right)\left(22 \ \frac{\frac{\text{ft}^3}{\text{min}}}{\text{person}}\right) + \frac{0.06 \ \frac{\text{ft}^3}{\text{min}}}{\text{ft}^2}\right)$$

$$= 7520 \text{ ft}^3/\text{min}$$

From the psychrometric chart, the specific volume of the dry air is approximately 14.7 ft³/lbm. The humidity ratios of the entering and exiting streams are 0.0292 lbm/lbm and 0.0026 lbm/lbm, respecitvely. [ASHRAE Psychrometric Chart No. 1 - Normal Temperature at Sea Level]

The mass flow rate of the condensate is

$$\dot{m}_w = \left(\frac{7520 \ \frac{\text{ft}^3}{\text{min}}}{14.7 \ \frac{\text{ft}^3}{\text{lbm}}}\right)\left(0.0292 \ \frac{\text{lbm}}{\text{lbm}} - 0.0026 \ \frac{\text{lbm}}{\text{lbm}}\right)\left(60 \ \frac{\text{min}}{\text{hr}}\right)$$

$$= 816.46 \text{ lbm/hr} \quad (800 \text{ lbm/hr})$$

The answer is (D).

76. Plotting the supply air condition, the outside condition, and the indoor design condition on a psychrometric chart, the humidity ratio for the supply air 0.012 lbm/lbm, for the outside air is 0.00085 lbm/lbm and for the indoor design conditions is 0.0084 lbm/lbm. [ASHRAE Psychrometric Chart No. 1 - Normal Temperature at Sea Level]

Find the rate of moisture that is needed by the building based on supply air conditions with a flow rate of 20,000 cfm. From a psychrometric chart, the specific volume of the supply air (105°F db, 25% rh) is 14.5 ft³/lbm dry air. [ASHRAE Psychrometric Chart No. 1 - Normal Temperature at Sea Level]

The total rate of moisture added by the supply air is

Moist-Air Cooling and Dehumidification

$$\dot{m}_w = \dot{m}_{DA}(W_{SA} - W_{IDC}) = \rho Q(W_{SA} - W_{IDC})$$

$$= \frac{Q}{v_{SA}}(W_{SA} - W_{IDC})$$

$$= \left(\frac{\left(20{,}000 \ \frac{\text{ft}^3}{\text{min}}\right)\left(60 \ \frac{\text{min}}{\text{hr}}\right)}{14.5 \ \frac{\text{ft}^3}{\text{lbm}}}\right)\left(0.012 \ \frac{\text{lbm}}{\text{lbm}} - 0.0084 \ \frac{\text{lbm}}{\text{lbm}}\right)$$

$$= 298 \text{ lbm/hr}$$

The outside air will supply only 4000 cfm. From a psychrometric chart, the specific volume of the outside air (35°F db and 20% rh) is 12.5 ft³/lbm dry air. [ASHRAE Psychrometric Chart No. 1 - Normal Temperature at Sea Level]

Find the needed increase in specific humidity to achieve the same humidification for the entire building.

$$\dot{m}_w = \frac{Q}{v_{OA}}(W_{DOAS} - W_{OA})$$

$$W_{DOAS} = \frac{\dot{m}_w v_{OA}}{Q} + W_{OA}$$

$$= \frac{\left(298 \ \frac{\text{lbm}}{\text{hr}}\right)\left(12.5 \ \frac{\text{ft}^3}{\text{lbm}}\right)}{\left(4000 \ \frac{\text{ft}^3}{\text{min}}\right)\left(60 \ \frac{\text{min}}{\text{hr}}\right)} + 0.00085 \ \frac{\text{lbm}}{\text{lbm}}$$

$$= 0.0164 \text{ lbm/lbm}$$

From the psychrometric chart, for a temperature of 75°F and a humidity ratio of about 0.0164 lbm/lbm, the relative humidity needed for the supply air for the dedicated outdoor air system (DOAS) is 88% (90%). [ASHRAE Psychrometric Chart No. 1 - Normal Temperature at Sea Level]

The answer is (D).

...mizer can save energy by using ...e building instead of recirculated ...air handler unit depends on the dif-...es between the supply air and the ...is reason, the economizer should be ...he enthalpy of the outside air is less ...py of the return air, which is the ...room design condition. (The refrigerat-...ature may have to be adjusted for a new ...at factor.)

The answer is (D).

...the heat loss for the copper and steel pipes ...le of values for heat loss from pipes. For a 2 in ...ipe at 120°F, the heat loss is 36.1 Btu/hr-ft. For ...steel pipe at 280°F the heat loss is 499 Btu/...Heat Loss from Bare Copper Tubing in Still Air ...F] [Heat Loss From Bare Steel Pipe in Still Air at ...

...total heat loss from both pipes is

$$q = \left(\left(36.1 \, \frac{\text{Btu}}{\text{hr-ft}}\right)(30 \text{ ft}) + \left(499 \, \frac{\text{Btu}}{\text{hr-ft}}\right)(20 \text{ ft})\right)\left(24 \, \frac{\text{hr}}{\text{day}}\right)$$

$$= 265{,}512 \text{ Btu/day} \quad (266{,}000 \text{ Btu/day})$$

The answer is (D).

79. Use the table of vibration isolation properties for various equipment types and locations found in *ASHRAE Handbook—HVAC Applications*. [Selection Guide for Vibration Isolation]

For cooling towers with rotational speeds of 301–500 rpm, the appropriate isolator type is a restrained spring isolator. [Selection Guide for Vibration Isolation]

The answer is (B).

80. ASHRAE Standard 62.1 requires 5 ft^3/min of outdoor air per person for an auditorium seating area, plus 0.06 ft^3/min per square foot based on the area of the auditorium. [Minimum Ventilation Rates in the Breathing Zone, Based on ANSI/ASHRAE Standard 62.1-2007]

According to the same table, the maximum estimated occupancy, is 150 people per 1000 ft^2.

The rate of ventilation required is

$$Q = \left(\left(\frac{150 \text{ persons}}{1000 \text{ ft}^2}\right)\left(\frac{5 \, \frac{\text{ft}^3}{\text{min}}}{\text{person}}\right) + \left(\frac{0.06 \, \frac{\text{ft}^3}{\text{min}}}{\text{ft}^2}\right)\right)(5000 \text{ ft}^2)$$

$$= 4050 \text{ ft}^3/\text{min} \quad (4{,}100 \text{ ft}^3/\text{min})$$

The answer is (B).